TEN BIRDS THAT CHANGED THE WORLD

TEN BIRDS
THAT CHANGED
THE WORLD

STEPHEN MOSS

First published by Guardian Faber in 2023

Guardian Faber is an imprint of Faber & Faber Ltd,
Bloomsbury House, 74–77 Great Russell Street,
London WCIB 3DA

Guardian is a registered trade mark of
Guardian News & Media Ltd,
Kings Place, 90 York Way, London NI 9GU

Typeset by Typo•glyphix, Burton-on-Trent, DEI4 3HE
Printed and bound by CPI Group (UK) Ltd, Croydon, CRO 4YY

A CIP record for this book
is available from the British Library

ISBN 978–1–783–35241–8

To my dear friends Lucy McRobert and Rob Lambert –
historians, environmentalists and birders – with grateful thanks.

CONTENTS

INTRODUCTION

Everyone likes birds. What wild creature is more accessible to our eyes and ears, as close to us and everyone in the world, as universal as a bird?

Sir David Attenborough

For the whole of human history, we have shared our world with birds.

We have hunted and domesticated them for food, fuel and feathers; placed them at the heart of our rituals, religions, myths and legends; poisoned, persecuted and often demonised them; and celebrated them in our music, art and poetry. Even today, despite a growing – and very worrying – disconnection between humanity and the rest of nature, birds continue to play an integral role in our lives.

Ten Birds That Changed the World tells the story of this long and eventful relationship, spanning the whole of human history, and featuring birds from all seven of the world's continents. It does so through those species whose lives, and their interactions with us, have – in one way or another – changed the course of human history.

But why birds? Why not mammals or moths, beetles or butterflies, spiders or snakes, or even domesticated animals such as horses, dogs or cats? All these, like birds, have been exploited and celebrated by us, and are central to our history and culture. Yet of all the world's wild creatures, birds are the one group with which we humans have had the closest, deepest and most complex relationship over time.

Partly this is down to their ubiquity. There is nowhere on the planet – from the poles to the equator – that you will not find birds. They are omnipresent, not just in space, but in time, too. During spring, summer, autumn and winter, you can see them; for much of the year, you can hear them too.

But this alone does not explain our fascination with birds. Along with other species – and indeed inanimate objects such as cars – we frequently anthropomorphise them, celebrating (and sometimes condemning) their supposedly human characteristics.[*] Throughout history, across very different cultures, we have found some birds cute and lovable, others aggressive and hateful, even though to them we are just another large, lumbering creature it is usually better to avoid.

For example, we often describe birdsong – and its positive effect on our mood – in musical terms, referring to the 'dawn chorus' or the 'orchestra'. We regard the peacock's display as 'putting on a show' for our benefit, or laugh at the comical antics of penguins. In the same breath, we might refer to birds of prey as 'ruthless killers', crows as 'devious' and vultures as 'loathsome scavengers' – conveniently overlooking the essential task they

---

[*] As the American author Polly Redford wisely noted, when we seek to impose human characteristics on wild creatures, we are bound to be disappointed: 'A bird is not noble any more than a dog is faithful, a pig greedy, a donkey stubborn, or a fox sly.' *Raccoons & Eagles: Two Views of American Wildlife* (New York: E. P. Dutton, 1965).

perform in cleaning up rotting meat and animal carcasses.

Our fascination with birds comes mainly from two aspects of their lives: their ability to fly, and their gift of song. Of these, the aspect we most envy is flight, as these lines from the Second World War poet and airman John Gillespie Magee demonstrate:

> Oh! I have slipped the surly bonds of Earth
> And danced the skies on laughter-silvered wings . . .[*]

The ability of birds to take to the air, and soar high into the skies – one so far removed from our own capabilities, and performed with such elegance and grace – sets them apart from our lowly, earthbound existence. It is a gift we have envied since prehistoric times, which only in the last two centuries or so – thanks first to the Montgolfier brothers' balloon, and then the Wright brothers' aeroplane – have we been able to emulate.[†]

Even today, when we are able to board a jet airliner and travel to the furthest corners of the globe, we continue to be enthralled by migratory birds' ability to do the very same journeys, finding their way to their destination and back without the aid of modern navigation systems.

---

[*] Fighter pilot Magee died in December 1941, aged just nineteen, when his Spitfire aircraft collided with another during a training exercise. His untimely death makes these words, from his sonnet 'High Flight', even more poignant.

[†] Ironically, two of the species featured in this book – the dodo and the emperor penguin – have lost the ability to fly. Originally this provided them with an advantage, enabling them to adapt and survive; subsequently, however, their inability to use the power of flight to avoid the malign acts of humanity has led to their actual or potential extinction.

Birdsong is in many ways even more central to our lives, having inspired musicians, poets and countless everyday listeners for thousands of years. Recently, scientists have discovered that one reason we are so fascinated by birdsong is that it really does lift our mood.[1] For the bird itself, on the other hand, the act of singing is a life-or-death struggle to repel rivals, attract a mate, and reproduce, and thereby pass on its genetic heritage to the next generation before its brief life is over.

A third reason why we are so fascinated by the lives of birds is that they share many of our own habits and behaviours. Indeed, at times, as the cultural historian and commentator Boria Sax notes, they behave in ways that look remarkably like an avian equivalent of human society.[2]

But does that mean that birds have influenced the course of human history, and even changed the world, as the title of my book suggests? I believe it does. The stories told here reveal the enormous influence that particular species or groups of bird have had on historical and current events, or on important aspects of our lives.

These range from a cumulative effect, over centuries, to specific events during a brief but critical period of human history. Birds have brought about social revolutions, changed the way we look at the world and, at certain tipping points in time, led to paradigm shifts. And the effects have been remarkably different, from the economic to the ecological.

Each of the ten birds I have chosen relates to a fundamental aspect of our humanity: mythology, communication, food and family, extinction, evolution, agriculture, conservation, politics, hubris and the climate emergency. All are interwoven with our close, continuing and ever-changing relationship with birds.

The narrative of *Ten Birds That Changed the World* is loosely chronological, with each of the ten chapters focusing on a single species (or group) through time, in roughly historical order.

From when Noah sent out the RAVEN from the Ark, birds have been at the heart of our superstitions, mythology and folklore. So I begin my story in prehistoric times, when this huge and fearsome member of the crow family featured in creation myths all the way across the northern hemisphere, from the First Nation Americans, via Norse culture, to the nomadic peoples of Siberia. Yet the raven's influence is not confined to the past; it continues to shape our view of the world today.

Roughly 10,000 years ago, soon after human beings began the shift from nomadic hunter-gathering to agriculture, and settled down to grow crops and raise livestock, they saw the major advantage to be gained in domesticating the wild birds that lived around them. One of these was a shy, cliff-dwelling species, the rock dove, originally bred for food, but later prized for its extraordinary ability to carry messages over long distances. Its descendant – the feral PIGEON – is now found throughout the world. Often denigrated, or simply ignored, this humble bird has helped win battles, and even change the course of the two world wars.

Domesticated birds provided not just food, but spiritual and social sustenance too. One of the most important examples is the WILD TURKEY of the Americas – still the centrepiece of Christmas feasts in Britain and Europe, and of Thanksgiving in

North America. Now produced on an industrial scale, the turkey is increasingly the subject of bitter argument about whether we have the right to exploit other living creatures for our own selfish ends.

The aftershocks and the human cost of European exploration and colonialism, from the fifteenth century onwards, are still being felt today. Of the many avian casualties of this period, none is more famous than the DODO. This huge, flightless relative of pigeons lived for many millennia on the oceanic island of Mauritius, yet it could not survive the seventeenth-century invasion of humans, and the various predatory animals they brought with them. Today, this icon of extinction can teach us useful lessons about our problematic relationship with endangered species, and how we might save them from the dodo's fate.

The rise of the new discipline of evolutionary science, during the eighteenth and nineteenth centuries, threatened to bring down the religious edifice on which civilised society had been built. The key turning point came in 1859, when Charles Darwin published *On the Origin of Species*, whose contents changed the very way we see the world around us. Yet, as we shall discover, it was not Darwin himself, but the scientists who followed in his footsteps, who eventually realised the importance of DARWIN'S FINCHES as a classic example of evolution in action.

It is often assumed that modern industrial agriculture began in the period following the Second World War. More than a century earlier, however, the droppings harvested from vast colonies of a South American seabird – the GUANAY CORMORANT – had provided the fertiliser needed to launch a boom in intensive farming.

This ultimately altered the landscape of North America and Europe for ever, marking the start of the long decline of farmland birds and wildlife, and changing the way we grow, consume and regard food.

Other species were under threat, too. The SNOWY EGRET of North America fell victim to the fashion trade, its ornate plumes adorning women's hats and dresses, driving this elegant waterbird to the brink of extinction. A backlash against such wanton cruelty – and the illegal killing of people who sought to protect the species – led to the formation of bird protection organisations, including the Audubon Societies in the USA, and the RSPB in Britain. Yet even today, brave men and women, working to save wild creatures and the places where they live around the globe, are being murdered for their cause.

The eagle has always been associated with the strength of nations and empires: first in its symbolic significance for the Ancient Greeks and Romans; then as the icon of the Holy Roman Empire, Germany and Russia; and finally, in the form of the BALD EAGLE, as the national bird of the USA. Yet eagles also have a darker history, as the emblem of totalitarian regimes: first in Nazi Germany, and now, in today's America, among supporters of the far-right. How this mighty bird came to represent the worst in human nature is one of the most disturbing stories in the book.

The story of a campaign against one small and once ubiquitous bird is, though, arguably even more shocking. Politicians have often fallen victim to hubris; none more than all-powerful despots. The story of China's Chairman Mao is a salutary one: he took on nature and lost. Mao's war against the humble TREE

SPARROW not only resulted in the species almost being wiped out, but also led ultimately to the deaths of tens of millions of his own people.

Finally, the fate of the EMPEROR PENGUIN – the only bird that breeds during the bitterly cold and harsh Antarctic winter – is now shared by the whole of humanity, as we career headlong towards the global climate crisis. Will the penguin's warning – through its rapid decline and impending extinction – come too late? Or will we, at one minute to midnight, manage to pull ourselves – and the rest of nature – back from the brink?

Never has the need to interrogate our relationship with the natural world been quite so urgent. Just during the course of my own lifetime, the number of birds on this planet has plummeted, thanks to a toxic combination of habitat loss, persecution, pollution and the climate emergency. There are now fewer than half the individual wild creatures – including birds – on Earth than there were in 1970;[3] the human population, meanwhile, has more than doubled, from 3.7 billion to 8 billion.[4]

Despite this dramatic decline, there remains a vestige of hope: the realisation that birds have never been more important to us, or to our continued future on the planet. We rely on them as we always have: not just for food, fuel and feathers, but also to enhance our understanding of the natural world, to which, owing to their ubiquity, they have become the most prominent gatekeepers.

There has never been a better time to focus on the long, tumultuous, and always fascinating relationship between birds and humanity, at the very moment when the current environmental crisis threatens to plunge us, and the natural world, into chaos and oblivion. It just might help us to forge a better relationship for the future.

Stephen Moss

Mark
Somerset
UK

RAVEN

Corvus corax

And he sent forth a raven, which went forth to and fro, until the waters were dried up from off the earth.

Genesis, chapter 8, verse 7

As dusk was falling on an early autumn day, a woman was working outside her home, in Boulder Canyon on the Colorado River. Yet she was finding it hard to focus on the task in hand. Close by, a large, black bird was uttering a constant chorus of loud, raucous cries.

The bird was a familiar one – a raven – but its behaviour that afternoon struck the woman as very odd. However much she tried to ignore it, the raven's calls were getting louder and more persistent. As she later recalled, 'It was putting on a fuss like crazy.'

In exasperation, she looked up, as the raven passed directly over her head. It landed on a nearby rock, just above where she was standing. Only then did she realise why the bird was behaving so strangely.

Among the rocks, barely twenty feet away, an animal was crouching: a cougar,* staring directly at her with its piercing yellow eyes. The beast – weighing over 50kg, more than the woman herself – was about to pounce. At less than five feet tall, she was about the size and weight of a deer, the cougar's usual prey. So if it did attack, she would at the very least be badly wounded; at worst, she would die.

The woman rapidly backed away from the cougar, calling out in fear. Her husband heard her panicked cries and arrived on the scene, scaring the predator away.

---

* *Puma concolor*, also known as the mountain lion or puma.

After she had recovered from the shock, the woman spoke about her narrow escape. She was in no doubt about what had happened: 'That raven saved my life.' The news media declared her survival to be little short of a miracle.[1]

But let us take a step back for a moment and focus not on the thoughts and feelings of the woman, but on the instincts and motives of the bird itself. Why would the raven want to warn her against a potentially fatal attack? And if there is no satisfactory answer to that question, what is really going on here?

Since prehistoric times, wolves and ravens have worked together to find food – sometimes co-operating with human hunters, at other times with mammalian predators. Ravens are too small to kill prey as large as a deer: something only wolves or humans – and cougars – can do.

But, compared with the raven, these large terrestrial mammals lack one major advantage: they cannot fly. Only the raven can reconnoitre a large area of ground, locate potential prey and then return to guide the hunters towards the target. If the hunters succeed in making a kill, they will feast on the animal's flesh. But when they are done, they leave the remains behind, with enough meat on the bones for the ravens to scavenge a hearty meal.

So while we might wish to see the raven's intentions as benign, isn't the opposite likely to be the case? Is it not far more probable that the raven was *intentionally* luring the cougar towards the woman, hoping that it would succeed in making a kill? Then both the cougar and the raven would have feasted to their heart's content. As the eminent ornithologist Bernd Heinrich, who recounted the tale in his book *Mind of the Raven*, notes, 'Everything I know about ravens . . . is congruent with the idea that ravens communicate not only with each other, but also with hunters, to get in on their spoils.'[2]

As an example of how we so often misunderstand the motives and actions of birds, it is hard to see how this story could be bettered. It teaches us an important lesson: that when it comes to wild creatures, we must take care not to assume that they are somehow 'on our side'. Sometimes they might be, but only if they also benefit from this temporary alliance.

The unvarnished truth is that, like every other bird in this book, ravens are simply thinking about themselves and their own survival. It is a truth we would do well to bear in mind.

We can tell that human beings and ravens have a long history together by examining the origin of the bird's name. 'Raven' is one of the oldest of all the names we use for birds, having first come into use long before the birth of Christ.[*]

We know this because, as with a handful of other bird names such as swallow and swan, the word for the raven is more or less the same in all the Scandinavian and Germanic languages – including English.[3] Therefore we can reasonably deduce that they all derive from the same original root, based, of course, on a humanised version of the bird's call.[4]

The word used in Icelandic to this day, *hrafn* (with the 'f' pronounced as 'v'), is likely to be the closest modern equivalent to

---

[*] The word goes back to at least the middle of the first millennium BC, to Proto-Germanic, an early branch of the Indo-European family tree of languages, and almost certainly much further than that.

the name uttered by our prehistoric ancestors, as they gazed up into a cold, grey sky and attempted to mimic the sound of this remarkable bird.

Ravens would have regularly followed human hunters – just as they would have tracked other large predators – in order to feed on the remains of their kills. But this was not a purely one-way transaction. In return, ravens would, as we have seen, alert humans and other mammalian predators to the presence of their victims.

This early, semi-symbiotic relationship with humans goes a long way to explaining why ravens feature so prominently in the mythologies of so many early cultures. Indeed, of all the world's birds, the raven is the one most central to the origin stories of ancient civilisations. Right across the northern hemisphere, from Alaska to Japan, via Britain and Ireland, Scandinavia and Siberia and the Middle East, the raven is not only the primary bird of myth and legend; it is also, in most cultures, the very first bird to be thus mythologised.

Many other birds have significant roles in global mythology. These include owls, noted for their supposed wisdom; cranes and peacocks, long recognised for their intricate courtship dances, especially in parts of Asia; the sacred ibis, linked to the religions of Ancient Egypt; eagles, which represent strength and power (see chapter 8); and the resplendent quetzal, one of the world's most beautiful and sought-after birds, which played a key role in the Pre-Columbian cultures of Central America. But significant though these all are, none have quite the same importance, geographical scope or historical longevity as our relationship with the raven.

Ravens also have a long and distinguished history as portentous messengers. This goes back at least as far as Ancient Greece, where Apollo (the god of prophecy) used them to deliver messages, although, as we shall see in other contexts, the birds were not always very reliable.[5] One of the best-known bird legends of all has it that if the resident ravens at the Tower of London ever leave, the United Kingdom and its monarchy will fall.[6]

And in case you imagine that in the modern world, the raven no longer has such a hold on our beliefs and cultures, consider this. When the American author George R. R. Martin came to choose a bird to symbolise the power of prophecy – and take messages, like a supercharged carrier pigeon – in his novels (and later the TV series) *A Game of Thrones*, there was only one possible contender: the raven.[7]

But why is the raven at the heart of so many ancient and modern mythologies? What is it about this particular member of the crow family that has singled it out, in so many eras and locations, and in such a diversity of cultures, for this crucial role? As with other birds that give rise to stories, myths and legends, it comes down to the character of the bird itself: the raven's habits, its behaviour and, above all, its intelligence.

Clever, resourceful, adaptable, crafty, opportunistic. Just five words of many that apply to ravens – and also, of course, to us. Like human beings, alongside whom they have lived for tens of thousands of years, ravens are able to change their behaviour to suit differing circumstances. Like human beings, they are able to solve problems, learn from their experiences and even vary what they do following a setback, so as to be more successful next time. And, just like human beings, they evoke a wide range of responses: from deep loathing to respect, admiration and even love.

But there is one other aspect of the raven's character that made it the ideal subject for such myths: its independence of spirit. We first notice this trait in one of the earliest stories about the bird, and indeed the first mention of any bird in the Bible: from the Old Testament Book of Genesis, the story of the Great Flood.

After forty days, Noah is desperately seeking dry land for the Ark. He decides to release two birds – the raven and the dove – and the first to be sent out is the raven.[8] The dove follows soon afterwards, but is unable to find a place to land and returns to the Ark. The raven is never seen again.

That independence – an unwillingness to bend to the will of its human counterparts – is a theme that can be found in virtually every story about ravens, in high and low culture, and from ancient times to the present day. The three great Abrahamic religions – Judaism, Christianity and Islam – share a belief that mankind is superior to all other creatures (for example, giving man dominion over them in the very first chapter of the Bible).[9] Yet the raven goes against this, doggedly refusing to be anything other than an equal partner – almost as if it considers itself as another human being.

For many observers, that's exactly how ravens are perceived. The nineteenth-century Scottish ornithologist William MacGillivray was not known for his sentimentality; the accounts in his epic, five-volume work *A History of British Birds* rarely stray from the purely scientific and descriptive. Yet when it came to the raven, even he could not resist the lure of anthropomorphism – the attribution of human characteristics to a non-human species:

> I know no British bird possessed of more estimable qualities
> than the Raven. His constitution is such as to enable him
> to brave the fury of the most violent tempests and to subsist
> amidst the most intense cold; he is strong enough to repel any
> bird of his own size, and his spirit is such as to induce him to
> attack even the eagle . . . in sagacity he is not excelled by any
> other species.[10]

MacGillivray might be writing admiringly about a fellow human being – perhaps a war hero or explorer – rather than a bird. This close identification of ravens with our own character – the best, and sometimes the worst, of our human traits – is also revealed by the ambiguity at the heart of so many myths and legends about the bird. Ravens can be good or evil; a powerful ally or a feared adversary; an unclean scavenger or a valued aid to keeping city streets clean.* They are often regarded as a symbol of hope, yet also (even simultaneously) as one of ill-omen. Whichever way we look at them, and however hard we try to pin them down, they remain an enigma.

It is these very human qualities – which, as we shall discover, are usually derived from the behaviour of the bird itself – that make

---

* In the Old Testament Book of Leviticus, ravens are condemned as 'an abomination' that were forbidden to be eaten – presumably because, like vultures, they are known to scavenge for their food and are therefore regarded as 'unclean'. Leviticus, chapter 11, verses 13–15.

the raven's enduring position in our lives so compelling. And by cementing the place of birds at the very centre of our culture, the raven ultimately changed the way we view the world.

What of the raven itself: the biological and ecological version, rather than the cultural, historical and mythological one?

The common raven[*] is (along with the thick-billed raven, found only in the Horn of Africa) the largest member of the crow family, *Corvidae*. These are by far the largest members of the order *Passeriformes* – also known as passerines or perching birds – which comprises 140 families and roughly 6,500 species, well over half of all the different kinds of birds in the world.[11]

Like other corvids, individual ravens vary considerably in size and weight.[†] They are also incredibly long-lived, especially compared with other passerines, which usually survive for just two or three years; even less for many smaller species. Ravens have been found to live as long as twenty-three years in the wild, though a more typical lifespan is between ten and fifteen years.[12]

For a concise yet evocative description of the raven, it is hard to beat this account from ornithologist Derek Ratcliffe's authoritative monograph of the species: 'In appearance it is a striking creature . . . with a heavy pick-axe bill. The apparent jet blackness

---

[*] *Corvus corax*, also known as the western or northern raven. '*Corax*' means 'croaker', a reference to the bird's deep and resonant voice.

[†] Typically, adult ravens are 60–70cm long, with a wingspan of 100–150cm and weighing 1.15–1.5kg – the male being, on average, larger and heavier than the female. To put this into perspective, a raven is roughly eight times longer and a whopping 300 times heavier than the world's smallest passerine, the pygmy bushtit of Java, and at least two-and-a-half times heavier than a carrion or American crow.

of plumage is enlivened in closer view by a glossy iridescence of purple, blue and green. In the air it shows an aquiline spread of wing, with splayed-finger primaries and a large, wedge-shaped tail.'[13] Ratcliffe, who spent many days in the field watching and studying ravens, also describes the bird's characteristic flight behaviour and unique call: 'In a more leisurely mood, the Raven . . . will frequently indulge in curious flight antics, rolling over onto its back and tumbling for a moment before reversing the movement. And, advertising its presence to all and sundry, it utters deep, resonant croaks and barks that carry far into the distance.'[14]

It is always hard to disentangle the biological and cultural aspects of the raven from one another, but when Ratcliffe describes the bird as 'the spirit of the wilds' he surely has both the physical and metaphorical characteristics of the bird in mind. To truly understand the raven's character, you really do need to see – and hear – the bird for yourself. Once you have experienced a close encounter with this fascinating creature, you will never mistake a raven for an ordinary crow again.

Just like human beings, ravens are incredibly successful: they can be found all the way across the northern hemisphere, including large swathes of Europe and Asia, and also much of North America, having crossed the land bridge between the Old and New Worlds several million years ago.[15] As a result, the raven has the largest range of all the world's 130-plus members of the crow family.

One key to the species' success is that ravens have been able to adapt to a wide variety of different climatic conditions, habitats and altitudes. The ornithologist Karel Voous noted that only the peregrine has managed to exploit a greater variety of environments,[16] while the editors of the standard work on the birds of the western Palearctic (Europe, North Africa and the Middle East) suggest that the raven is so adaptable that the concept of 'habitat' hardly applies.[17]

Ravens make their home from beyond the Arctic Circle to the deserts of North Africa; in hills and mountains, along coasts, in forests, on farmland and at the edge of cities; and from two-thirds of the way up Mount Everest to low-lying islands in the North Pacific.* In all these places, they have developed a close – though sometimes rather uneasy – connection with human beings. This relationship goes back many thousands of years, long before the beginnings of modern human civilisation.

As Derek Ratcliffe notes: 'The raven is . . . perhaps more intimately entangled with the cultural life of earlier peoples than any other bird in the whole of history'.[18] As we shall discover, this relationship reveals itself in some unusual and often surprising ways, some of which we are only just beginning to understand.

On 2 September 2009, an amateur archaeologist, Tommy Olesen, was excavating a site near the village of Lejre, in eastern Denmark, when he came across a tiny silver figurine, only eighteen millimetres high and weighing a mere nine grams. Two months later, it was revealed to the press and public at the nearby Roskilde Museum, where it remains on display today.

---

* Other species of raven can be found elsewhere in the Middle East, North America, across much of Africa, and Australia. They are absent from South America and Antarctica.

The object, which dates from roughly AD 900, is of a human figure on a throne, accompanied by two birds, one on either side. The identity of the figure, known as 'Odin of Lejre', has been the subject of dispute, but most experts believe it depicts the Norse god Odin, flanked by his two faithful ravens, Huginn and Muninn.[19]

Second only to Thor (whose fame has recently been boosted by his appearance in the Marvel Cinematic Universe movie franchise), Odin is one of the best-known characters in Norse mythology.[*] One-eyed and bearded, he was dubbed the 'father of the gods' and is celebrated for his wisdom, a quality attained through his close bond with that pair of ravens. Huginn means 'thought' and Muninn 'memory' or 'mind'.[20]

Every morning, according to legend, these two birds would fly all the way around the world, before returning to Odin's shoulders to whisper in his ears everything they had seen on their journey. This close relationship led Odin to be given the name *Ravneguden*, the Raven God.[21]

Scholars have long debated the symbolic meaning of Odin's ravens. Their ability to share their thoughts with him is linked by some to shamanism, in which the human protagonist connects with the spirit world by entering a trance-like state,[22] while the ravens may also represent a concept from Norse mythology known as *fylgja*, whose characteristics include shape-shifting between human and animal, good fortune and the concept of a guardian-spirit.[23]

---

[*] Odin features prominently in the mythology and folklore of all the Scandinavian and Germanic countries, including Anglo-Saxon England, while a version of his name, Woden, still survives in the name of the middle day of our week: Wednesday.

Both these ideas clearly influenced the plot of *Game of Thrones*, in which the paralysed boy Bran[*] regularly enters into a trance and appears to 'become' the three-eyed raven, enabling him to see the past, the present and (using a third eye) into the future.[†]

Its author George R. R. Martin has confirmed that when he put these birds at the centre of his story, he had Odin and the ravens in mind. He describes them as 'fearless, inquisitive, strong flyers . . . and large and fierce enough to make even the biggest hawk think twice about attacking them'. He also notes their supreme intelligence and concludes, 'Small wonder my maesters use them as messengers to tie the Seven Kingdoms together.'[24]

Along with the ravens, Odin is often depicted with two wolves, Geri and Freki. Again, these have been the subject of speculation as to their symbolism and meaning, but others see their presence as being rooted in the real world. Bernd Heinrich suggests a behavioural rather than symbolic explanation for the connection between man, wolf and raven in the Odin story.[25]

He proposes that this might reflect the real relationship between these three highly successful species: an early example of symbiosis, or co-operation for mutual benefit, between human hunters and the two wild creatures. As he points out, 'In a biological symbiosis

---

[*] A name deriving from a Celtic word for raven or crow.

[†] Martin calls this gift 'Greensight': defined as 'the ability to perceive future, past, or contemporary but distant events in dreams'. While the human characters in *Game of Thrones* die on a depressingly regular basis, the ravens are a constant presence virtually from page one to the very end of the five volumes published so far (and all eight TV series). Oddly, the sound of the raven in *Game of Thrones* is a strange, strangled call, far higher in pitch and much less resonant than the normal raven call. I would guess that it is a recording of a crow.

one organism typically shores up some weakness or deficiency of the other(s).' Being one-eyed, Odin needs help to see; he is also prone to being forgetful; hence the presence of the two ravens as his aides: 'He also had two wolves at his side, and the man/god–raven–wolf association was like one single organism in which the ravens were the eyes, mind, and memory, and the wolves the providers of meat and nourishment.'[26]

Heinrich goes on to discuss the origins of this relationship and the way it symbolises our loss of connection with the natural world. The Odin myth, he suggests, encapsulates the relationship between the humans and the other two creatures, in what he calls 'a powerful hunting alliance'.[27]

Over time, however, as human civilisation advanced, that bond between man, wolf and raven began to unravel. As our ancestors shifted from a nomadic, hunter-gathering existence to a settled life as farmers, so our relationship with the raven changed too: from a co-operative friend and ally, it became a competing foe and rival.*

That shift saw the first notable change in the complex and changing fortunes of the raven during the past few thousand years: from hero to villain – and, ultimately, back again.

Almost halfway around the world from Scandinavia, in the Pacific North-West region of North America, the raven also features

---

* Incidentally, it has been suggested that the word 'ravenous' derives from the greedy feeding habits of ravens: see Rachel Warren Chadd and Marianne Taylor, *Birds: Myth, Lore and Legend* (London: Bloomsbury, 2016). However, the *OED* proposes that the word has a completely different origin, though the English spelling (switched from 'ravinous' or 'ravynous' to ravenous) may have come about through a later (and coincidental) association with the name of the bird.

strongly in ancient culture and mythology. Like their counterparts in Europe, the people of the First Nations would have formed a close and symbiotic relationship with these birds, to help them find food.[28] It was then but a short step to incorporate the raven into their origin myths.[29]

In indigenous cultures like theirs, the raven is regarded as the creator of the world and the surrounding universe, including the sun and the moon. These stories share other common themes: ravens are shape-shifters, able to take on human and animal form; they keep secrets; and they offer valuable lessons to their human counterparts. Most importantly, ravens remain fiercely independent, always driven by the desire to satisfy their own needs rather than the needs of others. This, as we shall see, is central to the closeness of the relationship between human beings and ravens, throughout our shared history.[30]

Fire is often at the centre of these stories, with the raven's all-black plumage sometimes attributed to its feathers – originally white – being burned by the smoke from a firebrand it was carrying. Yet the bird's symbolism remains equivocal: while it represents the creation of the world, it is also a playful trickster – not qualities we usually associate with an all-powerful deity.[31]

Similar creation myths featuring the raven can be found across the Bering Strait in Kamchatka, Russia. Here, as in North America, the raven is often depicted as a trickster. The close connection between the two cultures is hardly surprising, given that the ancestors of North America's indigenous peoples originated in north-east Asia, having headed east into the Americas some 20,000 to 14,000 years ago.

Elsewhere, the raven (or one of its close relatives) is central to the culture and mythology of Ancient Greece, Rome, the Celtic civilisations, China, Japan, India, Australia and the Middle East: not only in the Bible, but also the Quran, in which a raven shows Cain how to bury his victim, his brother Abel.[32]

Ravens are not just a symbolic presence throughout human history, but a very real and tangible one. These huge black birds would have been a familiar sight on battlefields, from those of the Ancient Greeks and Romans to those of the Saxons, Vikings, Normans, right up to the fifteenth-century Wars of the Roses and the final bloody encounter fought on British soil, the Battle of Culloden in 1746. They were there for one reason only: to feed on the corpses of the dead and the bodies of the dying, often starting their feast by pecking out the soft parts such as the eyes. Because of this gruesome behaviour, they were understandably regarded as portents of death or disaster.[33] Edward A. Armstrong, in his 1958 book *The Folklore of Birds*, attaches much importance to this, devoting an entire chapter – portentously entitled 'The Bird of Doom and Deluge' – to the raven.[34] Yet as Armstrong accepts, like many birds in myth and folklore, the raven is not a one-dimensional symbol of evil, but often subtler and more ambivalent, able to take on a benevolent role as well as a malevolent one. Like its cousin the magpie, a raven may be a sign of good fortune as well as bad, as in this traditional ballad:

> To see one raven is lucky, 'tis true,
> But it's certain misfortune to light upon two
> And meeting with three is the devil![35]

27

Yet ravens can, at least temporarily, be habituated by humans. One frequent role of ravens, in many cultures, is as guides, both spiritual and real. For the seagoing Vikings, ravens had both a symbolic and practical importance. Warriors decorated their shields and banners with the image of a raven, whose association with death would, they believed, intimidate their enemies. They also used the birds to help them find land, as they navigated their ships across the North Sea to invade Britain.[36]

One Norseman, the ninth-century explorer Flóki Vilgerðarsson, headed even further from home, intending to reach Iceland (which had been accidentally discovered by his compatriot Naddodd some years earlier).[37] Legend relates that he took three ravens to help him find land. When he released the first, it flew back the way he had come, suggesting that there was still a long way to go. The second bird flew above the ship, but then returned, indicating that they were out in the open ocean, far from land. The third raven also rose high into the sky, but then headed north-west and never returned. Flóki realised that the land he sought was near and, following the bird, he reached Iceland. For these exploits he was renamed Hrafna-Flóki – 'Hrafna' meaning raven.[38]

Again, like so much raven mythology and folklore, this story is rooted in the raven's actual behaviour. The classical scholar and ornithologist Jeremy Mynott cites one story that suggests that ravens had a special ability to 'share news' of a battle to other birds, from far and wide: 'After a particularly grisly massacre at Pharsalus in Thessaly in 395 BC, ravens were said to have gathered there in great numbers, 'having deserted all their usual sites . . . suggesting that they had some sense through which they could communicate with one another' [the latter part of this is a quotation from Aristotle].[39]

Stories such as this, and indeed almost all human encounters with ravens, return to the one universally accepted characteristic of the bird from ancient times to the present day: its profound intelligence. As Bernd Heinrich points out, since the ancient Norse, and all the way to the pioneer of animal behaviour studies Konrad Lorenz, humans have considered the raven one of the world's cleverest birds.

Like other corvids, ravens have a larger brain relative to their size than most birds, enabling them to perform complex tasks that other species are unable to tackle. Rooks, for example, have demonstrated in both field and laboratory studies that they are especially good at problem-solving,[40] while the New Caledonian crow is not only able to make and use tools, but can also choose the right tool for a specific task, revealing an ability to plan for future events: something once considered to be unique to humans.[41]

Ravens, too, are very intelligent birds and, perhaps more importantly, are perceived as such by human observers, who commonly describe them as wise, cunning, alert and prophetic.[42] Yet this is not, as it might at first seem, an example of us imposing human qualities onto the birds: ravens really *are* highly intelligent. Indeed, one recent scientific study suggests that ravens match great apes in their ability to perform complex tasks.[43] Another study proposes that, like apes, ravens are able to delay gratification: deferring a reward in the present in order to gain a greater prize later on.[44] Many scientists also believe that ravens demonstrate 'theory of mind': the ability to understand and take into account the thoughts and mental states of others.[45]

As if to confirm that the Ancient Greeks and other early civilisations might have been correct in ascribing unexplained powers of communication to ravens, the linguist Derek Bickerton has suggested that ravens are one of only four animal groups (the others being human beings, ants and bees) to demonstrate 'displacement' – defined as the ability to communicate concepts that are not immediately present in space or time.[46] Just as honeybees perform their 'waggle dance' to indicate the direction and distance from their hive of a new source of nectar, so ravens foraging alone are able to convey the presence and location of a carcass to the rest of their cohort, at their evening roost. Hence, perhaps, the collective name for this species: a 'conspiracy' of ravens.

Both juvenile and adult ravens engage in what looks very like 'play': tumbling through the air alone or in pairs, or sliding down banks of snow.[47] I have often watched a lone raven as it suddenly turns upside-down in flight, before twisting back again moments later, for no obvious reason other than that it looks like fun.

So although there are some caveats to comparing ravens too closely with humans, when we witness such behaviour it is hard not to. Certainly, this intelligence, but also their lighter and darker qualities, inform their portrayal in both classic literature and popular culture.

Mention the part ravens play in literature and you are likely to elicit the immediate response, 'Quoth the Raven, "Nevermore".' The line comes from one of the best-known of all narrative poems,

'The Raven', written by the nineteenth-century American writer, Edgar Allan Poe.[48] Published in January 1845, less than five years before Poe died, it is a masterpiece of Gothic literature, whose popularity, even almost 200 years after it was published, has yet to wane; it features in almost every online list of the top ten poems of all time.[49]

With a curiously jaunty, yet strangely ominous, rhythm, 'The Raven' tells the story of a man who awakens 'upon a midnight dreary' to hear a tapping on his door. The visitor turns out to be a raven which, once it has been let into the room, will utter but a single word: 'Nevermore'. The largely one-sided dialogue between man and bird continues, with the man getting increasingly angry and disturbed, as he realises that the raven's constant repetition of the word reminds him of the loss of his lover, Lenore. Gradually he descends into madness, for which he blames the 'grim, ungainly, ghastly, gaunt and ominous' bird.

Just as in many ancient myths about the raven, Poe is primarily using the bird as a dramatic device: mirroring and in some ways hastening the man's mental breakdown. The same approach is repeated in other appearances of the raven in literature, almost always as some kind of ill-omen.

At the start of the final act of Shakespeare's *Julius Caesar*, for example, ravens, crows and kites fly over the doomed army of the assassins,

> . . . and downward look on us
> As we were sickly prey.

The imagery leaves us in no doubt that the birds are condemning the dreadful act of regicide.[50]

Once Shakespeare had found an appropriate symbol, he tended to re-work and re-use it. Thus, in *Macbeth*, Lady Macbeth pronounces that:

> The raven himself is hoarse
> That croaks the fatal entrance of Duncan
> Under my battlements.[51]

We can infer from this speech, so early on in the play, that Duncan's terrible fate is already sealed, long before he is actually murdered – the raven's call functioning as a sinister chorus to the human action. The raven also appears as a harbinger of doom in *Hamlet* – 'the croaking raven doth bellow for revenge'[52] – and *Othello*, in which the eponymous tragic hero laments that a bad memory returns to him 'as doth the raven o'er the infectious house'.[53] This would surely have been recognised by Shakespeare's audience as a topical reference to how ravens would gather over the homes of plague victims, attracted by the imminent opportunity to feast on a corpse.

Another famous literary link with ravens is more benign, involving pets rather than plague. Charles Dickens successively kept three pet ravens (all, bizarrely, named Grip) in his London home, where they terrorised the household dogs and the author's children.[54] They also featured in his 1841 novel *Barnaby Rudge*. One bird even travelled with Dickens across the Atlantic to Philadelphia, where they met Edgar Allan Poe himself – an encounter widely believed to have inspired Poe to compose his most famous poem.[55]

Poe's *The Raven* and references to the bird in other literary works, including Shakespeare, often focus on the raven's voice. Bernd Heinrich links this back to the story of Odin's ravens, who speak to him on their return from their daily travels. Yet he also points out a paradox: that although the raven's voice is the aspect of the bird's behaviour that has attracted most interest and comment, 'I'm convinced that there is nothing that we know less about.'[56] 'I feel I can detect a raven's surprise, happiness, bravado and self-aggrandisement from its voice and body language,' concludes Heinrich. 'I cannot identify such a range of emotions in a sparrow or in a hawk.'[57]

It might be easy to dismiss this as pure anthropomorphism, were it not for two things. First, Bernd Heinrich has watched and studied ravens for much of his eighty-plus years, and so if he says he can detect those emotions, we should take notice. And second, because he is not alone: anyone who has spent any time with ravens knows that their calls do indeed have a very human quality. This perhaps goes towards explaining the enduring popularity of Poe's poem, with that hypnotic repetition of the raven's refrain, 'Nevermore'.*

References to classical literature often appear in popular culture – sometimes overtly, but more usually covertly, with familiar stories rehashed as the basis for the plot of a wholly new book or movie.

---

* Ironically, although 'The Raven' turned Edgar Allan Poe into a household name, he made little money from what has become one of the most reproduced and widely quoted poems in history. Its legacy is still evident today in the name of the Baltimore Ravens American Football team, which was chosen in a poll of fans in 1996, partly as a nod to the local baseball team, the Baltimore Orioles, and also because Poe is buried in the city.

So it was not surprising when, in October 1990, the makers of the TV series *The Simpsons* decided to base part of their Halloween 'Treehouse of Horror' episode on Edgar Allan Poe's 'The Raven'. What was more unusual was that, in what was then still widely viewed as a cartoon for children, they chose to include the poem unabridged, in its entirety. Narrated by the gravel-voiced Hollywood actor James Earl Jones and featuring the various members of the Simpson family playing the characters, this is a tour de force of television; a fine example of the unification of classic and popular culture.[58] It also manages to be very funny, yet also surprisingly disturbing.

Many other modern literary works also feature ravens; again, often echoing existing myths and legends. In his bestselling 1937 children's fantasy novel *The Hobbit*,[59] J. R. R. Tolkien features two wise old ravens, Roäc and Carc. Both names are clearly onomato-poeic, and surely a reference to Huginn and Muninn.

The filmmaker Walt Disney seems to have had a particular liking for ravens. They appear in the earliest ever full-length animated feature film, *Snow White and the Seven Dwarfs* (released in 1937, the same year as *The Hobbit* was published); in the 1959 film *Sleeping Beauty* (with the name Diablo, Spanish for 'devil'); and in its 2014 remake *Maleficent* and sequel *Maleficent: Mistress of Evil* (2019), in which the raven is able to transform into human form (another nod to the link between ravens and shamanism). As in those ancient myths, Disney's ravens have very fluid characters: sometimes funny, sometimes crafty, sometimes sinister – and often all three at once.

The raven's complex and ever-changing character was also appre-ciated by Lewis Carroll (the pseudonym of the Victorian author and mathematician Charles Dodgson). In his bestselling 1865 work *Alice's Adventures in Wonderland*, the Mad Hatter poses a riddle for his audience at the tea party: 'Why is a raven like a writing desk?' Generations of readers have attempted to find an answer to the puzzle, with solutions ranging from the practical (both flap – the wings of the raven and the lid of the writing desk) via the conceptual (because they produce very few notes) to the surreal (Aldous Huxley's 'Because there is a "b" in both and an "n" in neither').[60]

Most commentators now believe that Carroll was deliberately teasing his audience – the riddle does not actually have an answer. Again, perhaps he chose the raven for the very reason that our view of it is – and always has been – shrouded in ambiguity.

How we view the raven today is to a great extent influenced by this parade of ravens down the centuries, from ancient myths and legends, via classical literature, to popular culture. But it is also inevitably shaped by the long history of this charismatic bird living alongside us, as hunting companion, useful scavenger and, at times, mortal foe.

We have already seen how early humans formed a close, symbiotic relationship with the raven, which would have accompanied them on their hunting trips and guided them towards a potential meal. But when our ancestors switched from being hunters to farmers, ravens soon turned from ally to enemy, as they now threatened people's livelihoods by consuming their precious grain.[61]

During and after the medieval period, across much of Europe, the raven's role – and how we regard the bird – shifted again. Alongside the red kite, it became a familiar, and largely welcome, scavenger in towns and cities; one of the 'clean-up squad' that kept the streets free from rotting animal corpses. In England, the raven was even protected by Royal Decree, as one late-fifteenth-century Venetian traveller noted to his surprise. 'Nor do they dislike what we so much abominate . . . The raven may croak at his pleasure, for no one cares for the omen; there is even a penalty attached to destroying them, as they say that they will keep the streets of the town free from all filth.'[62]

But from the early eighteenth century onwards, as our cities became more hygienic and the need for these birds declined, public opinion began to turn against both kites and ravens. Persecution, rather than protection, became the order of the day. The same happened in the wider countryside, where ravens had long survived by feeding on dead livestock and presumably, like rooks and crows today, gained a reputation for killing weak young animals. At the start of the nineteenth century the raven bred in virtually every county in Britain, but by the end of Queen Victoria's reign, in 1901, it had disappeared from many parts of the Midlands and East Anglia, though it still had a foothold in England's southernmost counties.[63]

Threatened with constant harassment and persecution, ravens were forced to take refuge in remote, upland sites. In Derek Ratcliffe's words, the raven became 'an outcast creature of the lonely places, the sheepwalks, moorlands, mountains and rugged seaboard'. As the nineteenth-century naturalist and hunter Abel Chapman noted, 'today the Raven epitomises the type of a vanished fauna,' alongside wolves, bears and wild boars.[64] It is no coincidence that Britain's uplands feature so many raven-based place names, notably

Ravenscraig, Ravenscar and Ravensdale (respectively meaning raven's crag, rock and river valley), all of which refer to features in the landscape where ravens would have lived.[65]

Yet even in these remote locations, they were still persecuted. From the seventeenth to the nineteenth century, a bounty of four old pennies per bird (roughly equivalent to £4 at today's values) was offered for each dead raven (the young as well as adults).[66] This practice was recalled by the poet William Wordsworth from his childhood: 'I recollect frequently seeing, when a boy [in the 1770s or 1780s], bunches of unfledged Ravens suspended from the churchyard of Hawkshead, for which a reward of so much a head was given to the adventurous destroyer.'[67]

In *Silent Fields*, Roger Lovegrove's forensically detailed history of the persecution of 'vermin' in Britain, he shows that for the next 200 years and more, ravens continued to be persecuted, poisoned and harassed. Indeed, in parts of Scotland, laws permitting the killing of ravens continued long after the 1954 Protection of Birds Act and were not finally repealed until 1981.[68]

Sheep farmers and bounty-hunters were not the raven's only nemesis. The rise of pheasant shoots during the eighteenth and nineteenth centuries, and the determination of gamekeepers to safeguard their precious birds, was effectively a death sentence for the species. Roger Lovegrove suggests that the raven's decline in lowland areas was almost entirely down to 'the almost universal and incessant persecution especially by shepherds and gamekeepers'.[69]

Illegal persecution of ravens (and other predators such as hen harriers and golden eagles) continues on many shooting estates today. Despite this, however, ravens are now expanding their range: after centuries of persecution, they have finally returned to their former haunts across lowland England. Ravens have even, symbolically, come back to breed on the White Cliffs of Dover.[70]

To put this comeback into perspective: when I began birding in the 1970s, to see my first raven I had to travel more than 300km, from my home in London to Snowdonia in North Wales. Thirty years later, when I moved to the Somerset Levels, I barely saw a raven in the first year or two I lived here. Now they are a common sight – and sound – across these marshy flatlands: during the first Covid-19 lockdown, in spring 2020, I came across no fewer than thirty-five of these huge birds in a single field, less than a mile from my home.

My own anecdotal evidence is backed up by figures from the regular *Atlas* surveys, carried out since the 1960s, by a legion of amateur birders on behalf of the British Trust for Ornithology (BTO). The map for the very first *Atlas* of breeding birds, covering the years 1968–72, corresponds almost exactly with upland and coastal areas in the north and west of Britain, with virtually no records from the lowlands to the east and south.[71] The *Winter Atlas*, covering the years 1981–4, shows very little change: both reveal that fewer than half the 10km squares surveyed (44 per cent) were occupied, during either the breeding or the winter season.[72]

Fast forward another quarter of a century or so, to the *Bird Atlas 2007–11* and the picture could hardly be more different. In the intervening decades, the raven's breeding range expanded by more than two-thirds, while its winter range almost doubled.[73] The map still shows gaps in England's eastern counties but, as the editors note, 'The raven is now as much of a bird of pastoral or mixed lowland farmland and forestry as it is of the uplands.'[74] Today, a decade or so since that most recent national survey, the raven breeds more or less throughout Britain.

The resurgence of the species has also led to a change in attitudes. In the raven's case, familiarity has bred not contempt but admiration. We have learned to celebrate the presence of this bird, to marvel at its acrobatic antics as it tumbles high overhead in the summer sky and not just hear, but really *feel*, that deep, croaking call, coming from somewhere in the firmament.

And, in an echo of those early myths about ravens, there is one final story that places the bird at the very centre of our civilisation. In this, ravens are not simply important, but seemingly crucial: as the ultimate guardians against the fall of the United Kingdom.

The Tower of London is not just part of the history of London, England and the United Kingdom; in many ways it exemplifies that history. Built – or rather begun – by William the Conqueror nearly a thousand years ago, it has dominated the story of the nation ever since. And although it is now dwarfed by the high-rise office buildings that surround it, the Tower remains one of the most impressive buildings not only in Britain, but anywhere in the world.

I first visited the Tower of London with my mother, back in the late 1960s. Arriving on a sunny September day, more than half a century later, I once again felt that awe and wonder I had first experienced as a child. This is where many English monarchs spent the night before they were crowned, where others were imprisoned, and where two – Henry VI and Edward V – were murdered. It is also where three Tudor queens – Anne Boleyn, Catherine Howard and Lady Jane Grey – along with 400 others, including Thomas More and Thomas Cromwell – were executed. Yet, despite this long and grisly history, nowadays the Tower is probably best known for two unique attractions: the Crown Jewels and the ravens.

Before I arrived, I had checked out the official Tower of London website, where I read how the ravens came to be here in the first place; and more importantly, the oft-quoted belief that, if the ravens are lost, perish, or fly away, the kingdom will fall.

The origin of this story, according to the website, is reputed to date back more than 350 years, to the reign of King Charles II. The King's official Astronomer Royal, John Flamsteed, is said to have complained that the loud noises made by the ravens flying around the Tower were interfering with his concentration. This, he claimed, made it harder for him to make important astronomical observations from what would have been, at the time, one of the tallest buildings in London: William the Conqueror's original White Tower. But when King Charles ordered that the ravens be destroyed, he was supposedly told that if they left the Tower, the monarchy would be doomed. Faced with such a warning, the King changed his mind and decreed that at least six ravens must remain there for ever.

At the time, the capital had just been through the twin traumas of a deadly outbreak of plague, in 1665, followed by the disastrous Great Fire of London the following year. Miraculously – and, indeed, almost immediately – the nation's fortunes began to improve. After the devastating schism of the English Civil War and the shocking execution of King Charles I, his son had now successfully restored the monarchy. The plague outbreak turned out to be the last, while even the Great Fire, regarded as catastrophic at the time, swept away the remnants of the medieval city and allowed Sir Christopher Wren to rebuild the capital for the modern age. All was well in the world, and if the continued presence of the ravens in the Tower could not quite take all the credit for these successes, these birds could still serve as a symbol of the stability of the nation. A heart-warming tale.

There's just one teensy-weensy problem. It is, as I discovered when I talked to Christopher Skaife, Yeoman Warder and, since 2011, the official Ravenmaster at the Tower of London, almost totally untrue.

When I had met Chris a few years earlier, at a book launch hosted by Hatchards, the celebrated central London bookshop, he had worn his splendid 'Beefeater' uniform, in black and scarlet, but on my visit to the Tower it was his day off, so instead he was wearing a rather colourful mauve T-shirt, appropriately sporting the logo of the Baltimore Ravens American Football team.

While writing his entertaining and informative memoir, *The Ravenmaster*,[75] Chris researched the 'fall of the kingdom' story. He discovered that, like so many tales about Britain's history and heritage, it is largely based on myth. Indeed, there is no evidence that ravens were kept at the Tower at all before 1883, when the story first appeared in a children's book.[76] But the suggestion that ravens have been present at the Tower of London for much longer than this might contain a grain of truth. Before and during Charles II's reign (1660–85), ravens would have been a common sight in the streets of London, and always on the lookout for flesh to scavenge. One story suggests that those wild ravens would have fed on the bodies of those executed at the Tower, including the ill-fated Anne Boleyn, beheaded here in May 1536. Another grisly, and oft-repeated, anecdote claims that ravens pecked Lady Jane Grey's eyes out of her severed head as it lay beneath the executioner's block – what Boria Sax calls 'a final, posthumous indignity'.[77]

As Chris told me, the adventurer, poet and statesman Sir Walter Ralegh, who was imprisoned in the Tower from 1603 until his execution in 1618, wrote to his gaoler Lord Cecil to complain about 'the confounded ravens', which were pecking at the flowers in his garden. But crucially, Ralegh doesn't say the ravens *of* the Tower, but merely those *at* the Tower, so we cannot definitively infer that these were the forerunners of the captive birds here today. Nevertheless, Chris believes, all the evidence suggests that the idea of the ravens preventing the fall of the kingdom is almost certainly a fairly modern myth. Over the years, this was repeated again and again, until it became the default explanation for the birds' presence.

Ultimately, though, Chris doesn't care whether the story is true or not. For him, it is what it *signifies* – that vital connection between birds, people and history – that really matters. As he points out, 'these stories, like the ghosts that walk the corridors at night, are part and parcel of the Tower of London's character and fabric. They're real – at least in people's minds.'

I casually asked him whether the captive ravens have at least been here continuously since that first definite record in 1883. His answer – and the hesitation that preceded it – surprised me. He went on to explain that during the Blitz two of the Tower's ravens died of fright from the terrible noise, leaving just three survivors. Two of these then turned on the third bird and killed it; an incident, he notes, 'that has been airbrushed from history'. To make things even worse, the remaining pair then flew away, so for several weeks there were no ravens at the Tower at all: something the authorities somehow managed to hide from the press and public. Fortunately, despite the imminent threat of a Nazi invasion, the kingdom did not fall, and, after replacement birds were brought to the Tower, ravens have lived and thrived here ever since. And, despite all the evidence to the contrary, the belief that the Tower's ravens safeguard the kingdom still resonates widely with the British public.

Chris clearly loves his job, and especially his daily interaction with these magnificent and very intelligent birds, which he describes in unashamedly human terms: 'They all have their individual personalities; I get to see their mood swings, their happiness and sadness – all the emotions that we have as humans. I don't think there's any other bird that has had such a close relationship with us, and for so long.' Yet as he pointed out to me, unlike other creatures such as horses and dogs, which we have ultimately domesticated, ravens have always remained wild and untamed: 'When I started this job, I thought I could control the ravens; after all, I am called the Ravenmaster. I soon learned that I don't control them – they control me – they do whatever they want to do.'

It confirms the conclusion that ravens are not only one of the most intelligent of the world's birds, but also the species most equal to us humans; indeed, on occasion, superior to us. 'They've used us from time immemorial,' says Chris, 'and they always will.'*

I asked him if there are any drawbacks to the job of Ravenmaster. He paused for a moment.

No matter where I am, they are continually on my mind, every moment of the day. I'm never away from what they're doing: whether I'm here at the Tower, at home, or on holiday, I can never stop worrying about them, which can be difficult for me and my family. But when I lose that, I'll know it is time to move on.

Having spent time with Chris, I can't see him ever losing his deep connection with his, the Tower's – and indeed the nation's – ravens. As we parted, he revealed to me that his ultimate dream is to travel the world, observing ravens in their many and varied environments and speaking to people for whom these birds are at the centre of their lives and cultures. I sincerely hope he gets to fulfil his ambition.

Just before I left the Tower, I got to take a closer look at the ravens themselves. From a few feet away, they are even larger – and more impressive – than I imagined. That glossy blue-black plumage, the huge bill and, more than anything, that

---

* Chris also observed that ravens are creatures of habit: 'They like their rituals. They like their pecking order. They like to know who's who and what's what.' He likens this need for order to human gangs, whose members won't cross an invisible line into someone else's territory. Sometimes, these borders – and the pecking order between individual birds – abruptly change, before settling down into a new equilibrium. Again, just like human society.

extraordinary call, confirmed my belief that this is a bird different from, and ultimately more inspiring than, any other species on the planet.

2

PIGEON

Columba livia domestica

It always seemed to me a sort of clever stupidity only to have one sort of talent – like a carrier pigeon.

George Eliot

On a cold, wet and miserable afternoon in February 1942, the crew of the Bristol Beaufort torpedo bomber had successfully completed the mission to drop their payload over German shipping off the coast of Norway. Now they were on their way back home, looking forward to a hot meal, much-needed drink and warm bed.

But the men's cheerful mood soon turned to despair, as they realised that the plane had been hit by anti-aircraft fire, and one of the engines was rapidly failing. Their only choice was to crash-land in the middle of the North Sea; as they did so, the aircraft broke up, throwing them into the icy waters. Miraculously, they did manage to scramble aboard their life raft – a small rubber dinghy – which now bobbed perilously up and down on the heavy swell.[1]

Just before the crash, the radio operator had managed to send a brief SOS message to their base at RAF Leuchars, on the east coast of Scotland. An air–sea search was launched, but unfortunately, the signal was so faint that the rescuers could not work out the exact position of the stricken crew, and so failed to find them. Freezing cold, soaked to the skin and already starting to suffer from the effects of hypothermia, the men appeared to be condemned to their fate.

But they still had one small chance of survival. This particular crew carried another passenger on their flight: a hen pigeon officially called

NEHU.40.NS.1, but later renamed Winkie. By a miracle, Winkie had not just survived the crash-landing, but her container had surfaced right next to the life raft. As dusk fell, with no time to waste, the crew released the pigeon, watching with a glimmer of hope as she headed away from them, and into the gathering gloom.

Soaked in salt water, and covered in oil from the crashed aircraft, Winkie flew straight home – a journey of almost 200km, at night, and in terrible weather. Sixteen hours later, as dawn broke, she arrived, exhausted, back at her loft in the Dundee suburb of Broughty Ferry.

Her owner, George Ross, immediately telephoned RAF Leuchars, who sent out a last-ditch mission to try to find the aircrew. The duty sergeant cleverly used the wind direction, and the time and location of that last, brief SOS message, to work out the position of the ditched plane. Remarkably, this worked, and the entire crew was saved.[2]

Later on, they held a special dinner to mark the occasion, with Winkie as the guest of honour. She also became the first pigeon to be awarded the prestigious Dickin Medal, the animal equivalent of the Congressional Medal of Honor or Victoria Cross. [3]

Winkie's miraculous flight is just one of many extraordinary feats of endurance and navigational skill demonstrated by pigeons and doves, which have lived alongside us perhaps for longer than any other bird. One of the earliest examples of this long relationship is also one of the best-known: the Old Testament story of Noah and his Ark.

As we have already seen, after forty long days and nights, with the flood waters still not abated, Noah first sent out a raven, but the bird never returned. So he tried again: this time with a dove.

But the dove found no rest for the sole of her foot, and she returned unto him into the ark, for the waters were on the face of the whole earth: then he put forth his hand, and took her, and pulled her in unto him into the ark.[4]

Seven days later, Noah tried yet again – and, later that evening, the dove returned once more. This time, though, she carried a fresh olive branch (or more accurately, a twig) in her beak, proving that the waters were at last in retreat, and land was nearby.

Always a cautious man, Noah waited another seven days, before releasing the dove for a third and final time. On this occasion, she did not return. The flood that had destroyed all life on Earth, bar the creatures Noah had managed to save, 'two-by-two', in his eponymous ark, was over at last.

The tale of Noah's Ark is one of the most enduring of all Biblical fables, perhaps because it contains the seeds of how humans can survive a terrible environmental catastrophe – a lesson we might do well to heed today.

But it also reveals a crucial contradiction in our relationship with the natural world, embodied in the nature of those two birds that Noah released from the Ark. While the raven goes its own way, and never returns, the dove comes back not just once, but twice. Whereas the raven is fickle and untrustworthy, the dove is dependable and steadfast. Where the raven is rebellious, the dove is dutiful and obedient.

As an example of the ambivalent relationship between humanity and nature – in which we are both part of the natural world, yet also seek to master it – the contrast could hardly be clearer. This is especially true in the opposing ways we have historically viewed the raven and the dove. The raven is a symbol of war, the dove of peace. The raven is mysterious, the dove familiar. While the raven is wild, the dove is tame. And ultimately one represents raw, the other domesticated, nature.[5]

But exactly when – and in particular *how* – did that long-standing link between the dove and domesticity begin? And how has this deep and complex relationship between man and bird defined and shaped human history, especially in the field of communication, from the beginning of recorded time, all the way to the present day?

When it comes to Biblical timelines, we rely on a combination of archaeology and educated guesswork. Scholars have frequently attempted to link the story of Noah's Ark to actual historical events, but the chronology remains tantalisingly uncertain.

Historians have suggested that a major flood almost certainly did occur in the region we now call the 'Middle East',[6] sometime between 2,000 and 10,000 years before the birth of Christ. This theory suggests that a rapid melting of the glaciers, following the end of the last Ice Age, generated a wall of water in the Mediterranean that inundated the neighbouring Black Sea. Archaeologists have found evidence, dating to roughly 5,000 BC, that indicates widespread flooding across 150,000 square km of land. That's an area larger than England, and roughly the same size as a modern nation now threatened by flooding, Bangladesh.[7]

Interestingly for our story, that date would fit closely with a key moment in the history of human civilisation: the first time that wild doves were domesticated. This is thought to have taken place sometime between 3,000 and 8,000 BC, in the same region – probably in Mesopotamia, the area of present-day Iraq and its neighbours, situated on the Tigris–Euphrates river system. This area, later known as the Fertile Crescent, is considered to be one of the earliest cradles of civilisation, where formerly nomadic hunter-gatherers first settled down to live in one place, and began to grow crops and raise animals for food.

Early on in their newly settled life, people would have captured wild birds; not just to kill and eat immediately, but to breed for food and other by-products such as feathers: a practice we now call domestication. Given the lack of hard evidence, there is considerable dispute as to exactly when this happened, and also about which species of bird was the first to be kept by humans in this way. Most historians believe that the red junglefowl of southern Asia – whose descendants are better known as the domestic chicken – is likely to have been the first wild bird to have been domesticated, some 8,000 years ago.[8] But this may not tell the whole story.

For the dove, the first written evidence of domestication appears later than this, on cuneiform tablets from Mesopotamia, estimated to be about 5,000 years old. Written records from other ancient civilisations, including Egypt, Greece and Rome, also detail the taming and keeping of wild doves. However, it is likely that these birds were being kept and bred in captivity far earlier than the available evidence shows; perhaps as long as 10,000 years ago.

The primary reason human beings would have kept any species of bird was for food, in the form of meat and eggs. That is why so many domesticated birds are large, plump and good to eat: notably chickens, geese, ducks, turkeys and (in a semi-feral form) swans.

Doves and pigeons, although much smaller, also make good eating; young birds – known as squabs – which were killed and eaten at around three or four weeks old, are especially tasty. There is ample evidence for this elsewhere in the Old Testament, notably the book of Leviticus, which contains several mentions of doves, both wild and tame, being offered to God as an atonement for sin, often as a convenient substitute for the costlier lamb.[9]

Meat was not the only useful commodity produced by domestic doves. Their feathers, plucked after killing the birds for food, would have been used to stuff cushions and pillows, while their dung was gathered, dried and used as fuel for fires.

Yet it was a very different aspect of these birds' lives – one perhaps discovered by accident – which ended up being far more important, making the dove stand apart from all the many other species of bird domesticated by humans. That is their ability to find their way home from hundreds, sometimes even thousands, of kilometres away; an ability still exploited by legions of pigeon fanciers. This is what places the humble pigeon at the heart of our story.

But of all the more than 300 different members of its family, why was the rock dove chosen to be domesticated in the first place?

'Dove' and 'pigeon' are words that, rather like 'swallow' and 'martin', are basically interchangeable. True, dove is conventionally used for the smaller, more delicate species, and pigeon for the larger, more robust ones but, given that the woodpigeon used to be known as the 'ring dove', and the rock dove is sometimes referred to as the 'rock pigeon', this is not a distinction worth making.

The rock dove is a medium-sized member of the family *Columbidae*, measuring between 31 and 34cm, and weighing around 300gm. It is roughly the same size as the stock dove of Europe, and slightly longer than (although twice as heavy as) the mourning dove, by far the commonest North American species.

A genuinely wild rock dove – though with so many feral birds interbreeding with their wild relatives, that term is becoming increasingly hard to define – is mostly bluish grey in colour. It has a darker grey head, breast and neck (revealing an iridescent purple or green sheen when it catches the light), and a dark tail and wingtips, with two black wing bars visible when the bird takes to the wing.

As such, it looks remarkably like many feral pigeons, though these can also display a bewildering range of shades and patterns, ranging from almost black, through greys, browns and buffs, to pale cream and pure white. This variety is a result of two factors: deliberately selective breeding to produce an aesthetically pleasing appearance, and random pairings between the descendants of once captive birds which at some point escaped into the wild. [*]

As the rock dove's name suggests, this is not a bird of fields, woods and forests, where most of its relatives can be found. Instead, it nests mainly on ledges amongst rocks and cliffs, or in

[*] Ironically, the ancestor of the feral pigeon is now at risk because of its descendants' success. A 2022 study reveals that populations of pure wild rock doves in Britain and Ireland are now in serious danger of disappearing, because of interbreeding with feral birds. The study also suggested that this may be happening throughout the rock dove's global range, threatening the species with ultimate extinction. Smith *et al*, 2022. 'Limited domestic introgression in a final refuge of the wild pigeon', *iScience*.

crevices and at the entrances to caves. This would have brought the species into close contact with early hominids, who might even have shared the same living quarters.*

Long before rock doves were properly domesticated, the chicks would have been regularly taken from their nests for food. Evidence for this relationship goes back at least 67,000 years – many millennia before the arrival of modern humans into Europe – to a cave in present-day Gibraltar. Here, over a period of thousands of years, Neanderthal peoples harvested wild rock doves. While this cannot be truly described as domestication, it does form a bridge between the hunting of wild birds and that more recent and formal practice.[10]

The rock dove's original range stretched from the rocky coasts of western and southern parts of Europe, through the deserts of North Africa and the Middle East, and eastwards across a broad swathe of southern Asia. This almost exactly matches the distribution of many early human civilisations, so it is hardly surprising that people living alongside these birds soon began to exploit them for their own benefit.

They could hardly have picked a better candidate. Rock doves are colonial, nesting together for safety against predators; they also breed throughout the four seasons, laying up to six clutches of eggs in a single calendar year. Because the chicks are fed on a special, high-energy substance known as 'crop milk', they are able to grow faster than any other baby birds, doubling their weight less than

---

* Wild rock doves continue to nest in these places, as confirmed during the 1940s, when amateur ornithologist John Lees discovered that those breeding in coastal caves in Cromarty, eastern Scotland, did so all year round. See John Lees, 'All the Year Breeding of the Rock-Dove', *British Birds*, vol. 39, 1946.

two days after hatching.[11] Crop milk is unique, in that parent birds can produce it regardless of what they eat, enabling them to feed on whatever is available, at any time of year, and convert it into this energy-rich substance.[12]

Domesticated rock doves were therefore able to lay eggs and raise chicks all year round, at frequent and regular intervals. This was a huge advantage over other species exploited by early humans, such as seabirds, which produce feast or famine: an abundant harvest during the spring and summer, but absolutely nothing for the rest of the year, while they are away on the open ocean.*

The colonial nature of the rock dove's life cycle would also have made it easier to keep. It was soon discovered that by building the birds a home, with entry holes and safe nesting places, and providing food in the form of seeds or grain, these free-flying birds could be encouraged to return each night.

Dovecotes – the name later given to these structures – are first known from Ancient Egypt and Iran, where the birds' droppings were also used in the tanning of leather and to make gunpowder. In the Negev Desert, in present-day Israel, pigeon dung was also used during the Byzantine period (between the fourth and fifteenth centuries AD) as a fertiliser to improve the soil.[13]

Other sites in Israel where ancient dovecotes have been found include Jericho, Masada and Jerusalem, along with Petra in

---

* As can be seen from the story of the 'bird people' of St Kilda, who had to adopt ingenious methods of air-drying seabirds, harvested during the spring and summer, in dry-stone structures known as 'cleits', to provide food for the autumn and winter, when the birds headed out onto the open ocean and so were no longer available. See BBC Four series *Birds Britannia*, episode 2, 'Seabirds' (first broadcast in October 2010).

neighbouring Jordan.[14] They were also found in Ancient Rome, and later throughout medieval Europe, where they became a status symbol for nobles, signifying wealth and power.[15]

It was inevitable that, soon after rock doves were first domesticated, some would fly away and return to a wild state. As the European conquest of the globe gained pace, doves were also taken on expeditions for food, and either deliberately or accidentally released into the wild, where these adaptable birds soon formed self-sustaining populations.

As a result, the feral (also known as the domestic, city, or street) pigeon has a far larger range than its wild ancestor. Today, it can be found across much of North and South America, Europe, Asia, Africa, and Australia, though it is absent from the two polar regions, large deserts such as the Sahara, and other natural or semi-natural habitats including the African jungles and savannah, and tropical rainforests.[16] Feral pigeons are most abundant in large cities, famously gathering at urban landmarks such as New York's Central Park, St Mark's Square in Venice and (until they were largely eradicated around the turn of the millennium) London's Trafalgar Square.

Pigeons often form large flocks, encouraged by people either giving or dropping food, and so can become a genuine pest or, as they have been described, 'the Most Hated Bird in America'.[17] Of all the many derogatory terms used to describe these birds, perhaps the best-known – and most insulting – is 'rats with wings', which first appeared in the *New York Times* in 1966.[18]

Yet, as we shall discover, although pigeons may often be widely loathed, they can also be deeply appreciated, valued and loved.

The rock dove's 'homing instinct' – its ability to navigate back to its loft or dovecote, and carry messages over long distances, far faster than human runners or riders – has led to the species being dubbed 'the original Internet'.[19]

The first specific records of these birds being used as messengers go back almost 5,000 years, to roughly 2,900 BC. In Ancient Egypt, one of the most sophisticated early civilisations, pigeons were released from incoming ships to give advance notice of the arrival of important visitors.[20]

A few centuries later, in Mesopotamia, King Sargon of Akkad also used the birds, but in a rather more sophisticated way. Each messenger sent out from his headquarters would carry a pigeon: if they were captured, they would release the bird, which on its arrival home would signal that the King needed to send a second messenger – ideally using a different route – to avoid being caught.[21]

Ancient Greece and Rome also made frequent use of homing pigeons. In 776 BC, the results of the early Olympic Games were sent by 'pigeon post'. During his conquest of Gaul, Julius Caesar used them to send messages between different parts of his army.[22]

It should be noted that pigeons are hardly unique in their homing habits. All migratory birds – especially those long-distance migrants that make a twice-annual journey across vast swathes of the globe – display this ability to a greater or lesser degree. Barn swallows, for instance, not only navigate the 10,000km route from northern Europe to Southern Africa, but also often find their way back to the exact location where they were born.[23]

To do so, these global travellers use a complex series of navigational tools and cues, including the Earth's magnetic field; the

Moon and stars (when travelling by night); polarised light (when flying by day); topographic features such as coastlines and mountain ranges; and, as they get closer to home, local landmarks such as rivers and even man-made features including railways and roads.[24]

In a 2004 study, scientists at Oxford University fitted pigeons with tiny devices to track their route from a release site to their home. They discovered that some birds faithfully followed the routes of major roads, much as a human driver or satnav system might, especially when planning the easiest (but not necessarily the shortest) route.[25] To the researchers' surprise, when the birds followed road systems, this sometimes added as much as 20 per cent to the length of their journey; despite this, they continued to do so. As the lead scientist Tim Guilford points out, though this might make their trip more demanding from a physical point of view, and use more energy, it would be easier mentally. As he wryly notes, even crows might not fly 'as the crow flies'.[26]

Side by side with these very practical uses, pigeons and doves also feature in the mythologies, religions and belief systems of ancient civilisations. However, this is less about their homing abilities, and more because of their legendary fidelity. As anyone who has spent time watching feral or domestic pigeons knows, their courtship display is complex, and often comical, yet also rather endearing.

First, the male approaches the female, strutting around her to attract her attention, while fluffing out his neck and chest feathers like a prizefighter, in order to look larger and more impressive. At this stage, she often affects a complete lack of interest, continuing to go about the business of feeding as if he were not there at all, or stopping to preen her feathers with her bill. This behaviour is known as 'displacement activity', and is not very different, biologically speaking, from the way we might subconsciously play with our hair, or scratch our heads, while talking to someone we are attracted to.[27]

Gradually, though, the female pigeon does begin to accept the male's attentions, and switches from ignoring his presence to actively encouraging him. They may now preen one another, in what to us looks very like a rather touching act of affection, before he finally mounts her to mate – a procedure that lasts for just a few seconds. After mating, the birds usually return to more typical behaviour, such as searching for food.[28]

It does not take a huge leap of the imagination to suppose that, early on in the history of pigeon domestication, observers would have made the connection between these very visible and intimate courtship activities and our own behaviour. From there, it was but a small step to view doves and pigeons as symbolic of those qualities we humans are encouraged to emulate: love, fidelity and monogamy.

One of the earliest accounts of pigeon behaviour, explicitly offered as an example to be followed by human couples, comes from the Roman naturalist and philosopher Pliny the Elder. In his masterwork *Naturalis Historia* (*Natural History*), published in AD 77 (just two years before the author perished at Pompeii during the eruption of Vesuvius), he wrote: 'Chastity is especially observed by [the pigeon], and promiscuous intercourse is a thing quite unknown. Although inhabiting a domicile in common with others, they will none of them violate the laws of conjugal fidelity.'[29]

Perhaps inevitably, this unashamedly anthropomorphic view led to pigeons and doves being incorporated into the symbolism and teaching of the three main religions of the Old World – Judaism, Christianity and Islam. This may also have been based on the pagan beliefs that pre-dated those Abrahamic faiths. Archaeologists have discovered carvings and figurines of pigeons at temples dedicated to various gods in Mesopotamia, while in Ancient Greece, doves were closely linked to the goddess of sexual love and beauty Aphrodite (the Roman goddess Venus), depictions of whom often show her surrounded by white doves.[30]

For the birds, however, there was a downside to all this attention. In 1100 BC, the Egyptian pharaoh Rameses III sacrificed 57,000 pigeons to the god Ammon at Thebes, and this practice continued at least until the birth of Christ. According to the Gospel of Luke, when Mary and Joseph bring the baby Jesus to Jerusalem, 'to present him to the Lord', they make a sacrificial offering of 'a pair of turtle doves, or two young pigeons'.[31] Similarly, the dove – now white, as a symbol of purity – often represents the Holy Spirit, such as when the newly baptised Jesus 'saw the Spirit of God descending like a dove, and lighting upon him'.[32]

Most of all, the white dove has long been a symbol of peace, a connection first developed in early Christianity, but since adopted by many secular cultures as well. In this guise, as in the story of Noah, the dove is often depicted as carrying an olive branch. There are many examples of this in the works of Pablo Picasso, who was so touched by the concept that, when his fourth child was born in 1949, he named her Paloma, from the Spanish word for dove.[33]

*

But communication is by far the most important aspect of the relationship between us and these birds, especially during times of war. Pigeons have been employed as messengers throughout the history of human conflict, from Hannibal to Genghis Khan, while the Crusaders cleverly used pigeons they had captured from their enemy to send false information to their opponents.[34] At the siege of Mutina (modern-day Modena, in Italy), in 43 BC, the Roman general Decimus Brutus managed to send messages out of the city by carrier pigeon, which resulted in reinforcements being sent to break the siege, and ultimately defeat his adversary Mark Antony.[35]

During the Napoleonic Wars of the early nineteenth century, the use of pigeons to carry messages led to one of the most enduring myths of the period. On 18 June 1815, a coalition of British and other north European forces, led by the Duke of Wellington, defeated the French Emperor Napoleon Bonaparte at the Battle of Waterloo in present-day Belgium. It was later reported that a homing pigeon belonging to the Rothschild banking dynasty had crossed the Channel, reaching London a full three days before any human messenger could break the vital news. This, it was claimed, allowed the head of the firm, Nathan Rothschild, to first sell his British government bonds as the price dropped with the fear of defeat, and then, just before news of victory finally emerged, buy them back at a much lower price – thus making a huge profit.[36]

It seems a pity to spoil a good story, but in reality, Rothschild received the news of Wellington's victory from a human, rather than avian, source.[37] Nevertheless, the romance of the pigeon story means that it continues to be repeated to this day in supposedly reliable media outlets.[38]

Less apocryphal – and all the more most extraordinary – is the story of communication using pigeons during the Siege of Paris. For more than four months, from September 1870 to January 1871, beleaguered Parisian citizens were trapped by the surrounding Prussian Army. They came up with all sorts of ingenious ways to carry messages in and out of the capital, including hot-air balloons. These, however, proved cumbersome, and vulnerable to the Prussian guns, so instead, they reverted to the tried and tested use of carrier pigeons.

In a revolutionary development, the besieged citizens used early photographic techniques to capture an image of the message, which they then reduced in size and printed on very thin film. This new technique, known as 'microphotography', allowed thousands of messages to be carried by a single pigeon.[39] When received, each missive could then be read by projecting and enlarging it onto a screen. During the course of the siege, over one million messages were sent in and out of Paris by this method.[40]

The heyday of the use of pigeons as messengers came in the twentieth century, during two of the most significant conflicts in the history of humanity: the First and Second World Wars. The achievements of these birds undoubtedly saved many thousands of lives, and helped change the ultimate course of those titanic struggles.

Visitors to the National Museum of American History, in Washington DC, come to see some of the most precious artefacts of that nation's past: including the original Star-Spangled Banner, the flag that inspired the USA's national anthem.[41]

Those museum-goers might also seek out a more modest object on display: a stuffed and mounted carrier pigeon named Cher Ami. Perched on his one remaining leg, and a little moth-eaten, he may not look very special. Yet his story is one of the most remarkable to emerge from the conflict of 1914–18. [*]

Cher Ami (French for 'dear friend') was one of 600 carrier pigeons used by the US Army Signal Corps during the First World War. This bird – along with the men from his company – was stationed near the city of Verdun in north-east France, close to the front line of the battle with the German forces. Cher Ami flew a dozen missions in all, the very last of which was on 4 October 1918, just over a month before the Germans finally surrendered.

On that fateful day, one battalion of the 77th Infantry Division – a ragtag group of 550 conscripted soldiers, mostly from New York City – had become separated from the rest of the US forces. The surviving men, commanded by Major Charles Whittlesey, were now in grave danger: not from the enemy, but from a barrage of friendly fire being dropped by their own colleagues. Not only were they cut off, and suffering horrific casualties, but they were also too far away from their headquarters to communicate their position by radio. In desperation, Major Whittlesey began releasing one pigeon after another, each carrying details of the battalion's location in a capsule attached to one of its legs. But every time a bird flew up into the sky, it was almost immediately shot down.

[*] There is some debate as to whether Cher Ami was male or female. At the time, it was regarded as male, but when it was stuffed to be put on display, was then thought to have been female. I have used the male pronoun throughout.

As the human casualties began to mount, the commander sent out his very last pigeon, Cher Ami, carrying a brief – and, under the circumstances, remarkably restrained – message: 'We are along the road parallel to 276.4. Our own artillery is dropping a barrage directly on us. For heaven's sake, stop it.'[42]

The chances of success were slim, to say the least. And indeed, Cher Ami was shot during his flight, sustaining wounds to the chest, and losing his right leg and the sight of one eye. Yet he still made it back to his loft, and successfully delivered the capsule containing the vital message. Within hours, the 194 survivors from what later became known as the 'Lost Battalion' had been found and rescued, all thanks to one pigeon's instinctive determination to get back home.

Cher Ami finally returned to the US the following spring, as a military hero, having been awarded the Croix de Guerre with Palm – an award given for exceptional valour during battle – by a grateful French government.[43] Less than a year later, in June 1919, he died of the wounds he had received on that extraordinary flight.[44] General John J. Pershing, the commander of the American Expeditionary Force on the Western Front, paid a special tribute to Cher Ami: 'There isn't anything the United States can do too much for this bird.'[45]

In 2020, more than a century after his death, the pigeon's story was turned into a novel, *Cher Ami and Major Whittlesey*, by the author Kathleen Rooney. Narrated by the pigeon itself, it poignantly suggests that his incredible feats have been forgotten in the intervening years: 'I myself have become a monument, a feathered statue inside a glass case. In life I was both a pigeon and a soldier. In death I am a piece of mediocre taxidermy, collecting dust.'[46]

Cher Ami's story is indeed remarkable. Yet it is just one of countless examples of the exploits of tens of thousands of pigeons during the two world wars. When it came to sending life-or-death messages in battle, pigeons often trumped any other method of communication.

During the Second World War, even though wireless radio had been developed as a replacement for the ineffective landlines used in the previous conflict, carrier pigeons were still used extensively by both sides. The Signal Pigeon Corps (officially the US Army Pigeon Service)[47] consisted of 3,150 men and no fewer than 54,000 pigeons, each trained to carry messages – of which more than 90 per cent were successfully received.

On the other side of the Atlantic, the British Army Pigeon Service, which had been disbanded at the end of the First World War, was hurriedly relaunched in February 1939 (seven months before the outbreak of the Second World War), as the National Pigeon Service (NPS).[48] The NPS had the powers to effectively conscript Britain's one-and-a-half-million racing pigeons – via their 100,000-plus owners – to serve if required. King George VI immediately donated the birds he kept at his loft at Sandringham in Norfolk, but other owners were reluctant to hand theirs over – partly because, unlike the royal enthusiasts, the vast majority of pigeon fanciers were working-class and not at all well-off, so could ill afford to lose their valuable birds. Indeed, pigeons were even dubbed 'the poor man's racehorse'.[49] Nevertheless, over time, many fanciers did offer up – and in many cases sacrifice – their birds to help the war effort.

In late 1940, Operation Columba was launched by military intelligence. This was an ambitious scheme to drop homing pigeons, supplied by the NPS, into continental Europe, using tiny parachutes. Each carried messages designed to elicit information about what was going on in territories occupied by the Nazis, from the French Resistance and any other groups or individuals hostile to the invading force. This was highly dangerous – and not just for the birds. If anyone was discovered to have an unregistered pigeon in their possession, they could be arrested and shot.

For four years or so, from April 1941 onwards, more than 16,500 pigeons were parachuted into occupied countries. Just 1,850 – roughly one in nine – returned home, but that was more than enough to make the mission worthwhile. This extraordinary story – and the wider use of pigeons during the Second World War – is told in a 2018 book, *Secret Pigeon Service*, by the BBC Security Correspondent Gordon Corera.[50]

In a remarkable twist, worthy of the espionage author John Le Carré, the British intelligence service even decided to use pigeons as spies, aiming to infiltrate the German Army Pigeon Service. But how, if they didn't know the specific ways in which the Germans used rings to identify their birds, or the design of the capsules which carried the messages?

Then came a breakthrough. The British managed to capture two German pigeons in the North Sea, enabling them to copy the rings and capsules, attach these to their own birds, and then clandestinely release these 'undercover operatives' into occupied

Europe. In a counter-intuitive yet clever move, the pigeons they chose were not especially good fliers. This meant that, instead of immediately heading home, they tended to join up with local birds, insinuating their way into cohorts of pigeons being used by the Germans themselves. These fifth-columnists were then given messages by the Germans and released into England, eventually returning to their own lofts, allowing the British to read the secret German communications.[51]

Meanwhile, just as in the First World War, pigeons were also being used to carry messages during emergencies. One US Army bird, GI Joe, is credited with saving a group of more than 1,000 British troops, along with the Italian inhabitants of Calvi Vecchia, a rural commune north of Naples. He managed to bring back a message that the village had been successfully taken by the British infantrymen, and so prevent it being bombed by their own side from the air. Yet again, when radio technology had failed, a pigeon had succeeded.

For his efforts – he had flown 32km in just twenty minutes, arriving just as the aircraft were about to take off – GI Joe was, like Winkie, awarded the Dickin Medal, for 'the most outstanding flight made by a United States Army pigeon in World War Two'. Indeed, of more than fifty Dickin Medals awarded during the Second World War, no fewer than thirty-two were given to pigeons.[52]

Commando, another pigeon awarded the Dickin Medal, played an important role in helping the French resistance movement against the occupying German forces. Bred in Sussex by keen pigeon fancier and First World War veteran Sid Moon, Commando made almost a hundred flights from Britain to France and back.

In three of these, during the period from June to September 1942, he brought back vital information about the location of enemy troops and injured British soldiers. In 2004, more than six decades after he carried out his daring feats, Sidney Moon's granddaughter auctioned Commando's medal for £9,200.[53]

However, the most important part pigeons played in shaping the future course of the war was during the D-Day landings in June 1944, when the Allied forces launched a major invasion of continental Europe, via the coast of Normandy, by land, sea and air.

The generals organising the invasion faced a major problem. The usual means of communication – radio – might easily have been intercepted, giving away the time and place of the invasion to the Germans, so could not be widely used. Once again, pigeons came to the rescue. One bird, named Gustav, flew 240km into a strong headwind, from Normandy to the south coast of England, to deliver the vital news that the first troops had successfully landed on the beaches, and the invasion was under way.[54] Another pigeon, aptly named the Duke of Normandy, carried a further crucial message across the English Channel to Allied HQ, to let them know that a German defensive battery had been destroyed. His journey home took twenty-seven hours, through heavy winds and rain.[55]

Gustav's flight was later fictionalised in an animated British movie, *Valiant*, released in 2005, with voiceover contributions from, among others, John Cleese, Ewan McGregor and Hugh Laurie. Sadly, Gustav's death was not quite as heroic as his life: while cleaning out his loft, a nameless person accidentally trod on him.[56]

*

As well as bad weather – and the constant danger of being shot down by German marksmen – pigeons faced other dangers on their journey back home. One threat particularly enraged the British High Command: the possibility that their precious birds might be killed by peregrine falcons.[57]

The peregrine is the fastest creature on the planet, reaching speeds when hunting of up to 390 kph (over 240 mph). They specialise in one type of prey – birds – and although they have been known to target and kill a huge variety of different species, by far their commonest prey item is the pigeon.[58]

Pigeons are no pushover, even for such a fearsome predator. They may not be able to fly as fast as their assailant, but they are more manoeuvrable, able to twist and turn in mid-air to evade attack. Nevertheless, they do often get caught. So when it became known that pigeons carrying messages vital to the war effort were being killed by these powerful birds of prey, some of which, it turned out, had actually been trained and released by the Germans to target British birds, something clearly had to be done.

In 1940, a branch of the domestic security service MI5 named the Falcon Destruction Unit was tasked with ensuring that pigeon casualties were kept to a minimum. Five crack marksmen were trained and then, like the fictional spy James Bond, issued with a 'licence to kill'. Soon afterwards, this was made official, with the publication, by the Secretary of State for Air, of the Destruction of Peregrine Falcons Order.[59]

Driving an open-top American Packard car – the glamour of which was perhaps slightly spoiled by the caravan in which they slept being towed behind them – the squad toured the south coast of Britain searching for peregrines. Once they found the birds, they shot them, killing more than 600 during the course of the war. This methodical slaughter of peregrines had a devastating effect on an already falling population, reducing the nesting population in England by as much as half, and pushing this majestic raptor to the brink of extinction in southern Britain.[60]

Ultimately the exercise proved counterproductive to the war effort, because by then the Nazis had installed spies in the UK who were using their own pigeons to send secret messages back to Germany. By killing the peregrines, the British were thus unwittingly allowing these German pigeons – and the vital information they carried – to get through to the enemy. In a bizarre twist, another arm of the British government then began to use their own peregrines to target the German pigeons. It was a total failure: of just seven pigeons killed, all turned out to be British.[61] This might be regarded as comical, were the consequences of their incompetence not so serious, for pigeons, peregrines and people.

During the years after the Second World War, the UK peregrine population, already severely depleted by the actions of the Falcon Destruction Unit, was hit again by the widespread use of the agricultural pesticide DDT. This rose up the food chain and accumulated in the birds' system, causing their eggshells to become thin and fragile, and fail to hatch. As a result, numbers plummeted, and only really began to recover towards the turn of the millennium, more than half a century later.[62] The pigeons had suffered too: during the two world wars many homing pigeons perished from predation or exhaustion.

Lost but, like the human dead, not forgotten. Along with the awarding of the Dickin Medal to a handful of birds, their collective feats were formally recognised in November 2004 with the unveiling of a memorial in London's Park Lane, to all the birds and animals killed during wartime service. Officially opened by the Princess Royal, following a national appeal for funding, which raised £2 million, the Animals in War memorial has two inscriptions. The first is a simple dedication; the second, a brief but effective statement: 'They had no choice.'*

That statement is meant, of course, to suggest that the pigeons acted out of valour and a conscious determination to succeed in their mission. In fact, just like the raven, they were simply following their instinct, with no awareness at all of the human cost if they should fail.

In today's world, with such rapid advances in technology since the two world wars, we might imagine pigeons would no longer have an active role to play in the safe delivery of vital messages. But even into the twenty-first century there are instances when they have persisted longer than we might expect.

It was not until 2006 that the pigeon post in the east Indian state of Odisha, which carried daily communications between 400 police stations, was finally disbanded, as email had made the need for physical communication obsolete. Before then, the birds had

---

* Another, far less grand, memorial can be seen in the public gardens, near the seafront, in the Sussex town of Worthing. Easy to miss, the stone is covered with moss and the inscription is damaged and faded. It reads: IN MEMORY OF WARRIOR BIRDS WHO GAVE THEIR LIVES IN ACTIVE SERVICE.

saved thousands of lives during two natural disasters: a devastating cyclone in 1971, and heavy flooding in 1982.[63]

Consider the advantages of these birds over the hi-tech alternatives. Pigeons are able to fly rapidly, during the day or night, and find their way back to a specific place in a relatively short time. They do not carry anything that would enable them to be detected among other birds so, unlike radio signals, they cannot easily be intercepted. And, unlike human messengers, they cannot be interrogated by the enemy, nor betray their paymasters by acting as double agents.

In our increasingly paranoid world, though, even pigeons can sometimes be regarded as suspicious. So when, in 2010, a teenage boy in an Indian village near the Pakistan border came across a pigeon carrying a message written in Urdu, the local police detained the bird. They X-rayed it and, although they found nothing amiss, officially reported it as a 'suspected spy'. The incident led to a series of memes circulating on social media, mocking the officials' perceived paranoia.[64]

But perhaps they were right to be cautious. In May 2016, Fox News reported that the Islamist terrorist group ISIS was using carrier pigeons to send messages to undercover operatives beyond the borders of its so-called 'caliphate' inside Iraq.[65] A year earlier, reports had emerged from eastern Iraq of ISIS fighters rounding up fifteen young men and boys for the 'crime' of keeping and breeding pigeons. For this seemingly innocent hobby, relatives reported, some of those young men were executed.[66]

What of the future? With the advent of high-tech military drones, have pigeons now been rendered obsolete in the fields of war, espionage and organised crime? According to Dr Frank Blazich, a curator at the National Museum of American History (home to Cher Ami), the answer is an unequivocal no. Dr Blazich points to the many advantages of a reliable, low-tech, organic messenger, even in a world where new, cutting-edge technology rules. As he explains, thanks to large capacity microSD memory cards, a single pigeon can easily transport large files of video, sound or still images, over long distances, in a way that – unlike most drone technology – remains undetectable to conventional surveillance systems.[67]

The relative speed of pigeons and the Internet was tested out in 2009, when a homing pigeon in South Africa carried a memory stick on a one-hour journey between Howick to Durban, a distance of just over 100km. At the same time, the data was sent over a broadband link between the two sites. To the delight of pigeon fanciers everywhere, the bird beat the Internet. Now that broadband speeds have improved exponentially, if the race were repeated today, that result would most likely change, but the South African experiment exemplifies the enduring efficiency of carrier pigeons.[68]

Technology has its problems too. In August 2021, the British cyber-security expert Professor Alan Woodward complained that 'pesky pigeons' were perching on his newly installed satellite dish and causing his broadband service to drop out temporarily. The reason, he suspected, was that the grey, upturned dish on his kitchen roof looked rather like a bird bath. Or maybe the pigeons were simply getting their own back.[69]

Though pigeons such as Cher Ami and the Duke of Normandy rightly earned acclaim for their wartime feats, it seems our memories are rather short and selective. The presence of feral pigeons in our cities' most public spaces has, at times, led to a hostility bordering on paranoia, both from the authorities and from the general public. Over the years, pigeons have been subject to a wide range of eradication and deterrent methods, including patrolling hawks and falcons, shooting, trapping, electrocution and the removal of their eggs. This is despite studies showing that the population soon recovers from such depredations, simply because the birds are able to reproduce so rapidly.

The reason pigeons thrive in cities is because our urban spaces provide everything these adaptable birds need. They can obtain food and water, and find plenty of places to roost and nest, in what scientists call an 'analogue habitat': the artificial equivalent of natural cliffs, crags and caves.*

When they are not being actively vilified, city pigeons tend to be largely ignored, even by birders, who do not really consider them a 'proper' species. One notable exception was the ornithologist, author and broadcaster Eric Simms, who in 1979 published *The Public Life of the Street Pigeon*.[70] Living in the north-west London suburbs, Simms had plenty of opportunities to make close observations of these neglected birds, and did a fine job rehabilitating them in the minds of ornithologists, who previously had not considered feral pigeons a suitable subject for study.

Simms opens the book with an account of a Tube journey he made in early 1965, during which he and his fellow commuters observed a street pigeon board the train at one station (where

---

* The same also applies, of course, to the pigeons' greatest adversary, the peregrine, which explains why they too have moved into our cities during the past few decades.

the Tube is still above ground) and then alight at the next one (also above ground), seemingly oblivious to the attention of the passengers.[71] A decade later, he watched another pigeon, this time underground, feeding on a platform at Marble Arch Station in central London. He goes on to explain the behaviour of the street pigeon in great depth, so that, by the end of the book, the reader can only be astonished at what we have learned about this often overlooked, yet always fascinating, bird – notably its ability to adapt to living more or less disregarded alongside human beings.

The presence of pigeons in cities – and the problems they cause – is far from a modern phenomenon. Pigeons were recorded fouling the streets of Ancient Rome, while towards the end of the fourteenth century the Bishop of St Paul's Cathedral in London complained about people throwing stones at the pigeons and breaking the cathedral's windows.[72]

From that time onwards, as cities grew, so did their pigeon populations. However, in the early twentieth century, the replacement of horse-drawn vehicles by motorised ones (and the consequent loss of the grain on which horses were fed, which made up much of the birds' diet) did lead to a temporary fall in their numbers.

Following the Second World War, the urban pigeon bounced back; probably because of the wider availability of food, either dropped accidentally, or given deliberately, by city dwellers and tourists. The practice is famously celebrated in the 1964 movie *Mary Poppins*, in the song 'Feed the Birds'. The film's eponymous protagonist, an Oscar-winning performance by big-screen debutante Julie Andrews, sings touchingly of an old lady who sits on the steps of St Paul's Cathedral selling crumbs of bread for 'tuppence a bag'.[73] In a sanitised version of reality, this attracts flocks of white doves, rather than the more authentic dirty grey of the genuine London variety. More than half a century later, in a sign of the authorities' continued ambiguity towards feral pigeons, the cathedral advertises on its website a commemorative 'Feed the birds' cup and saucer, while rather inconsistently displaying a stern 'Please don't feed the birds' sign outside.[74]

St Paul's Cathedral is not the only London tourist venue to prohibit the feeding of pigeons. Another famous landmark and public meeting-place, Trafalgar Square, was once the best-known place in Britain to feed pigeons, with entrepreneurial seed-sellers hawking their wares for a lot more than 'tuppence a bag'.[*] Then, in 2003, the then Mayor of London, Ken Livingstone, made the radical decision to evict the seed-sellers, and use a team of four North American Harris hawks to scare away the birds. Why? The thousands of pigeons gathering in the square were now deemed a hazard to human health, because of their copious droppings (a typical bird produces about 12kg of faeces every year).[75]

The move proved unpopular on several counts. For Londoners and visitors to the capital alike, Trafalgar Square was synonymous with feeding the pigeons. Meanwhile, a row soon broke out about the cost of the scheme: an eye-watering £136,000: almost £30 per bird. The Mayor's office countered by pointing out that the square's statues,

---

[*] Trafalgar Square was so closely associated with pigeons that in the 1965 spy film *The Ipcress File*, starring Michael Caine as the anti-hero and spy Harry Palmer, his boss, Colonel Ross, feeds the pigeons from his office window ledge, against the backdrop of Nelson's Column.

including the lions around the fountain and Nelson's Column itself, would no longer need to be cleaned so frequently, potentially saving far more than the initial cost of evicting the pigeons.[76]

The protests eventually died down, and in September 2007 a total ban was imposed on feeding pigeons in Trafalgar Square and its surroundings, with penalties including a £500 fine for ignoring the rule.[77] Since then, many other European and North American cities have followed suit, with bans in, among others, Venice and New York.[78]

One frequent justification for this policy is that pigeons can carry a range of diseases, some of them potentially injurious – and in rare and very unfortunate cases even fatal – to human beings.[79] But not everyone agrees with the harsh way pigeons have been treated. Colin Jerolmack, professor of sociology and environmental studies at New York University, and author of *The Global Pigeon*,[80] has questioned local governments' and businesses' systematic demonisation of urban pigeons as 'pests', despite these birds having lived alongside human beings in our towns and cities for many thousands of years.

He traces the change in attitudes to the early 1960s, when pigeons seemingly lost their association with peace and holiness and became what he calls 'menacing vermin to be exterminated'. Jerolmack goes on to suggest that by eradicating the pigeons from London's best-known public space, the authorities have promoted the outdated, erroneous yet remarkably persistent view that cities are for people, not nature.[81]

Nevertheless, there is no doubt that in extreme and unusual cases, pigeons can pose a serious threat to human health. Supporters of bans on feeding point to a tragic incident in 2019 when a child died from a fungal infection that had probably been caught from pigeon droppings, owing to poor hygiene in a Glasgow hospital.[82] Whether that warrants an all-out war against city pigeons is, however, arguable.

I cannot leave the subject of urban pigeons without a reference to the greatest satirical songwriter of all time: Tom Lehrer, and his song 'Poisoning Pigeons in the Park', which foreshadowed the more negative attitudes towards these ubiquitous birds. In 1959, Lehrer released a new album, *An Evening Wasted with Tom Lehrer*,[83] which included this brief song, whose jaunty tone and rapid-fire delivery conceal some gloriously dark – and truly shocking – humour.

In reality, Lehrer was sympathetic to these popular urban birds; his barbed and witty lyrics were aimed at the US Fish and Wildlife Service, whose operatives at the time really were lacing corn with strychnine to kill pigeons in New York's Central Park.

Persecuted, poisoned and demonised pigeons may be, but they continue to exert a hold over our collective imagination, even when we may not realise it. Take a closer look at the universally recognisable symbol of the social media giant Twitter. This clearly represents a bird in flight – but which one?

That perhaps depends on which version you are looking at. Sometimes the logo is depicted as a blue bird against a white background; on other occasions it appears as a white bird against blue. The blue version has often been linked with a particular species: the mountain bluebird *Sialia currucoides*. This sparrow-sized, migratory member of the thrush family breeds in upland areas of western North America from Alaska to California – not all that far from Twitter's global headquarters in San Francisco.[84]

But in Twitter's white-against-blue version, it surely represents another species: one whose symbolic relationship with us goes back to the very beginnings of human civilisation. This would certainly fit in with what the former creative director of Twitter said, when an updated version of the logo was launched in June 2012: that the bird represents 'the ultimate representation of freedom, hope and limitless possibility'.[85]

What else could that be, but the most effective messenger of all the world's birds, the humble pigeon?

3

TURKEY

Meleagris gallopavo

Turkey: A large bird whose flesh, when eaten on certain religious anniversaries, has the peculiar property of attesting piety and gratitude.

Ambrose Bierce, *The Cynic's Word Book*, 1906[1]

Like any other American family, they sat down together on a chilly November day to eat and, just as importantly, give thanks to the Lord for that year's bountiful harvest. After saying grace, they tucked into a feast – at the centre of which was a huge, roasted bird.

The diners at the event – which has entered American mythology as the 'First Thanksgiving' meal – included the fifty or so surviving pioneers from the Pilgrim Fathers or Pilgrims, as they had become known, who, one year earlier, had travelled on the Mayflower *to the New World.*

These men, women and children had set sail from the Devon port of Plymouth on the morning of 16 September 1620.* Even with the advantage of what was optimistically described as a 'prosperous wind', the journey was a long, uncomfortable and often dangerous one. The voyage, which had been expected to be over in a few weeks, took two whole months, much of which was spent drifting without sails, because of the high winds.[2] They finally sighted land on 19 November; two days later, the Mayflower *docked at Cape Cod, in present-day Massachusetts.*

The Pilgrims might have safely arrived in the New World, but that first winter was horrifically hard: the bitterly cold weather meant

---

* All dates are given in the new calendar – at the time, they would have been ten days earlier.

they were unable to find a suitable site to build a settlement, so they remained living on the ship, where many perished from diseases such as scurvy, pneumonia and tuberculosis.[3] The ground was frozen, so that no crops could be planted; heavy snow storms prevented them from exploring the area around their new home. By the spring, just fifty-three settlers – barely half the original cohort – were still alive.[4]

From that low point, however, the Pilgrims' situation gradually began to improve. With the help of the indigenous Wampanoag peoples, they learned to survive in this harsh and unfamiliar land and, by the anniversary of their arrival in November 1621, they were finally able to offer thanks to God for their first successful harvest. For three days, they and the Wampanoag sat down together and feasted.[5]

This was the forerunner of the modern-day Thanksgiving, which, following a tradition established by President Abraham Lincoln in 1863, takes place in late November each year. As one of the Pilgrims at the feast, Edward Winslow, wrote in a letter back home to England: 'Our harvest being gotten in, our governor sent four men on fowling, that so we might after a special manner rejoice together, after we had gathered the fruits of our labors.'[6]

We cannot be certain what kinds of 'fowl' were cooked and eaten at that historic event. But given the abundance and size of one particular species, it is quite likely that the centrepiece of the feast was the largest game bird in the world: the wild turkey.*

---

* It has been argued that 'fowling' suggests that the main quarry would have been ducks or geese; however, such is the power of the idea of the 'First Thanksgiving', that such quibbles are usually forgotten. Incidentally, at that time 'fowl' would have referred to any large bird, including the turkey.

An estimated 250 to 300 million turkeys are consumed by Americans each year; of those, one in six is cooked and eaten for Thanksgiving. According to the National Turkey Federation, on what is dubbed 'Turkey Day', nearly 90 per cent of Americans feast on the bird, although this is as much for practical as cultural reasons: no other bird is large enough to feed so many people at a single sitting. Its presence at the heart of the feast has been described as 'the single most important role of the turkey in American life'.[7]

Statesmen from Benjamin Franklin to Alexander Hamilton have praised the turkey's ubiquity on this annual celebration, with Hamilton going so far as to proclaim that 'No citizen of the United States should refrain from turkey on Thanksgiving Day.' That includes the person at the very top: for well over a century, a live turkey has been presented to the US President, for him and his family to eat at Thanksgiving. However, in recent years the bird has been granted a presidential pardon and allowed to live out the rest of its days in peace.

In his detailed chronicle of the history of wild turkeys in the United States, published in 1914, Albert Hazen Wright concludes that the turkey was at the very heart – and stomach – of the United States' early history: 'Thus, we see how essential the wild turkey was to the explorer, how prominent a part of the larder it proved for the early pioneers and Indians, what sport it furnished our natives, settlers and foreign sportsmen, and how early it was singled out as our token of festival joy.'[8]

Later historians have cast doubt on the idea that the feast of 1621 has any real connection with modern Thanksgiving, point-ing out that the link between Thanksgiving and the Pilgrims did not become explicit until the American Civil War, more than two centuries after the original event is supposed to have taken place. That was when President Lincoln announced that the final Thursday in November would, from then on, be a national day of thanksgiving.[9]

But ultimately, as with so many stories in this book, the power of the myth has proved more potent than the truth. As one historian notes, 'Turkey is consumed at Thanksgiving feasts because it was a native to America, and because it is a symbol of the bounteous riches of the wilderness.'[10] Symbolism, it seems, trumps historical accuracy: as the famous line from the 1962 film *The Man Who Shot Liberty Valance* has it, 'When the legend becomes fact, print the legend.'

Today, the turkey – long since domesticated, and now produced in industrial quantities – is served not just at Thanksgiving, but also as the main meal at Christmas, on both sides of the Atlantic.[11] At these celebrations, as we sit together around a table, food serves to bring us together, enabling us to forget our usual family disagreements and quarrels. As the US writer and filmmaker Nora Ephron notes, 'The turkey. The sweet potatoes. The stuffing. The pumpkin pie. Is there anything else we can agree so vehemently about? I don't think so.'[12]

How the turkey came to represent the crucial connection between food, family and the concept of 'home' – and become what historian Andrew Smith calls 'an American icon'[13] – is an absorbing story, dating back long before the Pilgrims to the earliest American civilisations. It is also, against all the odds, as important in the modern age as it was to the two groups of people who shaped the history of that continent: the Native Americans and the European settlers.

After all, without the meat provided by the abundant and easily obtainable wild turkeys to those pioneering Pilgrims, it is possible that the Europeans would not have survived to colonise North America at all. So, had it not been for the turkey, the course of the world's history might have been very different.

The wild turkey – the ancestor of the domesticated version we consume at those annual festivals – is by far the largest of almost twenty species of game bird native to North America, and one of the largest of all the continent's birds. It is also the biggest and heaviest of almost 300 species in the order *Galliformes*, colloquially known as gamebirds.*

The *Galliformes* – which also includes pheasants, partridges, quails, grouse, peafowl and the junglefowls (ancestors of the domestic chicken) – is one of the world's oldest bird groups. The wild turkey is thought to have evolved roughly twenty million years ago; its ancestors would have lived alongside the dinosaurs.[14] Its closest relative is the ocellated turkey, found in the Yucatán Peninsula (south-eastern Mexico, and parts of Guatemala and Belize) – indeed, the Maya called the Yucatán 'Land of the turkey'.[15]

At first sight, a wild turkey looks remarkably like a slimmer, fitter version of its domestic counterpart. It is huge and bulky, with

---

* A big male turkey can reach a length of 125cm, and weigh 11kg. Females are far smaller and lighter: typically, between 76 and 95cm long, and weighing between 2.5 and 5.4kg – less than half that of their mate. The heaviest known male wild turkey, according to the National Wild Turkey Federation, weighed a whopping 16.85kg, making it one of the heaviest wild birds ever recorded.

a plump body, long neck and long tail, which the male fans out like a peacock when engaging in courtship display. The coloration appears mainly dark, though on closer inspection the subtlety of its many different shades becomes clear. These include iridescent greens and browns, whitish feather tips, an overall coppery sheen, and pale bluish or reddish (and largely featherless) head, neck and throat.

Turkeys prefer to live in mature, mixed forests and orchards, with some open grassy areas, but also thrive in marshes and even suburban areas: anywhere they can forage for food and find safe hiding-places from predators.* When threatened, unlike the ridiculously overweight birds we raise for food, a wild turkey can fly fast and low to get away. It can also run surprisingly quickly: up to 25 mph over short distances.[16] When they need to, turkeys can even swim, spreading out their tails, tucking in their wings and paddling along with their powerful feet.[17]

Wild turkeys are found across much of North America, though in some areas they are descended from birds that have over time been reintroduced for shooting. It is also likely that most of the turkeys found in the eastern half of the USA contain at least some genetic material derived from interbreeding with domesticated birds.

The species has also been introduced to various locations outside the Americas, with varying degrees of success. Small numbers have been successfully introduced to Germany and the Czech Republic

---

* Their extensive catalogue of enemies includes foxes, racoons, owls, crows, hawks and eagles, snakes, cougars, coyotes, bobcats, lynx, dogs and cats, most of which mainly target eggs and chicks rather than the larger and more feisty adult birds. One unfortunate turkey was even caught, presumably while drinking, by an opportunistic alligator.

(though some of these may be descended from domestic birds), while there are larger, self-sustaining populations in Hawaii and New Zealand.[18] During the mid-eighteenth century, King George II released several thousand turkeys into the royal park at Richmond, south-west of London, to be hunted with dogs and shot. The experiment ultimately failed, because so many birds were taken by poachers.[19]

In North America itself, the wild turkey population is estimated at roughly 6.7 million individuals (up from just 1.5 million in 1973).[20] This, however, pales into insignificance compared with the number of domestic turkeys, which, allowing for peaks and troughs around Thanksgiving and Christmas, can reach a phenomenal 420 million birds.[21]

Such a large and abundant bird was an obvious, easy and very welcome source of food for the early peoples of North America. However, as with other tasty species of gamebirds and waterfowl – including the ancestors of the domestic chicken, geese and ducks – the Native Americans must have eventually realised that life would be a lot simpler if they could keep and raise these birds in captivity. Doing so would be much easier than hunting them in the wild, and provide a reliable, year-round supply of meat and feathers.

Two Old World species, the chicken and the goose, are thought to have been domesticated between 7,000 and 5,000 years ago, with the pigeon following soon afterwards (see chapter 2).[22] Turkeys were domesticated later on: roughly 2,000–2,300 years ago, just before the birth of Christ. Until that time, the people of the Americas were mostly nomadic hunter-gatherers; it was only once they established permanent settlements, and could grow grain to feed themselves and their livestock, that they could begin to keep animals. This makes the turkey the only significant species to have been domesticated in the Americas (other species, such as the goose and chicken, were first domesticated in the Old World, before being brought over to the New).[23]

Domestication appears to have happened in two widely separated parts of the turkey's range, more or less simultaneously. It also involved two different peoples: one, in southern Mexico, who kept turkeys sometime before the Aztecs (who are often mistakenly credited as being the first to do so);[24] the other, the Anasazi, who lived in the Four Corners region of the south-west United States.[25] The Anasazi turkeys died out along with their civilisation, some 800 years ago, which means that the southern Mexican subspecies is the ancestor of all domesticated turkeys today.[26] That is ironic, given that this particular race is now critically endangered in the wild.

Evidence that domesticated turkeys were served and eaten at ceremonial events more than 2,300 years ago has recently been unearthed by archaeologists at the Florida Museum of Natural History in Gainesville. They discovered that turkeys were the centrepiece of feasts held by Maya rulers as long ago as 350 BC, roughly 1,000 years earlier than previously thought.

This revelation has made historians look again at the Maya civilisation. As the lead researcher Erin Kennedy Thornton notes, rather than regarding the early Mayans as subsistence hunters, we now understand that they had easy access to a plentiful food resource, which they could both control and manage:

'Plant and animal domestication suggests a much more complex relationship between humans and the environment – you're intentionally modifying it and controlling it.'[27]

However, food was not necessarily the main reason for domesticating turkeys. In 2012, archaeologists excavating a Native American village in Colorado discovered a mass grave of turkey bones, not discarded haphazardly after a feast, but carefully arranged and buried in what must surely have been some kind of ceremonial ritual.[28]

Some peoples, including the Cheyenne, believed that eating turkey would make them cowardly; instead, they kept the birds for their long tail feathers, which were used for arrows, to make clothing and blankets, or for ceremonial purposes. The bird's wings were made into makeshift brooms, or used as fans, while its spurs – the long, sharp claws on the back of the bird's legs – were sometimes made into earrings or fashioned into arrow tips to kill small animals.[29]

Turkeys were fed precious supplies of maize, to fatten them up and keep them alive, even when people went hungry after a poor harvest.[30] However, sometime during the Pueblo period (from AD 750–1500) people began to cook and eat those previously symbolic turkeys – possibly because they were short of food and had no other choice. Another reason may be because of threats from outsiders, which would have made it too dangerous to go on long hunts for other game animals. This can be inferred from research at archaeological sites: at roughly the same time as the bones of deer begin to vanish, turkey bones are found instead. Whatever the reason, as one archaeologist notes, 'there's a pretty dramatic shift from it being almost taboo to eat a turkey, to the wholesale raising and eating of birds.' Even today, the turkey still has important symbolic and cultural value among the descendants of those early hunters.[31]

To the south, in Mexico, the Aztecs consumed vast quantities of domesticated turkeys. In the early fifteenth century, Montezuma II (also known as Moctezuma) fed huge numbers of turkeys to his collection of large raptors – some sources suggest as many as 500 birds a day. When the culinary needs of his royal household are considered, it is possible that Montezuma's requirements exceeded 1,000 birds a day.[32]

For almost two millennia after they were originally bred for religious rituals and ceremonial feasts, small numbers of domestic turkeys would have coexisted with their wild counterparts, with a limited trade in the birds being carried out within the Americas. Then came the historical event – or series of events – that changed the history and culture of the Americas for ever: the arrival of the European conquistadores: first Christopher Columbus and Juan Ponce de Léon, then Hernán Cortés, and finally, a century after Columbus landed, the British explorer, poet and adventurer Sir Walter Ralegh.

It is possible that Columbus was the first European to set eyes on – and eat – a turkey: in 1502 he is known to have feasted on a large, unidentified game bird on an island off the coast of Honduras.[33] We know for sure that soon after the Spaniards' discovery of Mexico in 1518, they brought live turkeys back to Europe – a significant historical event in itself. As A. W. Schorger points out in his definitive history of the species, 'When the Spanish explorers brought the turkey to Spain . . . North America made its most important avian contribution to the history of the globe.'[34]

From Spain, the custom of keeping and eating turkeys spread rapidly northwards, through France, eventually reaching Britain by the middle of the sixteenth century. The man often credited with bringing the birds to England is the explorer and landowner William Strickland, whose family crest depicts a turkey cock, tail raised in display, supporting the Holy Bible.[35]

Unlike some other New World foods, such as maize, tomatoes, potatoes and chocolate, all of which took a long time to be accepted and consumed by the rather fussy Europeans, the turkey soon became a favourite, especially for the British. Indeed, it eventually displaced the goose, and another large, edible bird, the swan, whose flesh (especially of the adult birds) is notoriously tough.

The food historian Professor David Gentilcore suggests that because the turkey's white flesh was similar to the meats already eaten on that side of the Atlantic, people immediately understood how to cook and serve the bird. There was also, he believes, an element of snobbery. Meat – and especially expensive, imported meat – could be bought and consumed only by the higher classes of society; thus, the turkey soon became a symbol of status and wealth.[36] 'On a continent where fine dining still included eating storks, herons, and bustards,' one commentator archly notes, 'the meaty, succulent turkey was a sensation.'[37]

Over the next century or so, according to Andrew Smith, 'turkey farms proliferated throughout England.'[38] He notes that by the mid-seventeenth century, vast flocks of these birds were regularly being driven many miles to market, like cattle. The novelist Daniel Defoe, who published several volumes about his travels through his homeland, wrote of turkeys filling the roads between East Anglia and London, each flock numbering as many as 1,000 birds.[39]

Turkeys were so widely farmed that their meat soon lost its exclusive status, and became affordable for the newly emergent middle classes. Even before the Pilgrims arrived in North America, British cookbooks featured recipes for turkey.[40] It is quite possible that the arrival of these birds prevented widespread famine in sixteenth-century Europe. At the time, suggests Smith, farmland was becoming increasingly degraded, leading to severe shortages of staple foods: 'Europe was on the verge of widespread undernourishment and potential starvation. Too useful to remain the province of wealthy aristocrats, the turkey quickly became a widely available food for all but the poorest of the poor. In protein-starved sixteenth-century Europe, bigger was better.'

The custom of serving a turkey, boiled, roasted or braised, on Christmas Day appears to have begun soon after the bird's arrival in Britain, as early as the 1570s. However, with the rise of the Puritans, the very thought of celebrating Christmas went temporarily out of fashion, and at one point the tradition was effectively banned, not to return until 1660 following the Restoration.[41] By 1792, turkeys were so closely associated with the festive feast that the poet John Gay could coin the couplet,

> From the low peasant to the lord
> The Turkey smokes on every board.[42]

By the middle of the nineteenth century – the time of Charles Dickens – turkey had become standard fare at Christmas. Dickens himself may have helped enshrine the bird's place at the centre of Britain's festive meal – and, indeed, effectively invent the rituals of our modern-day Christmas. In *A Christmas Carol*, published in 1843, the repentant miser Ebenezer Scrooge sends his worker Bob Cratchit a turkey 'twice the size of Tiny Tim [Cratchit's small and sickly son]' for the family to enjoy a splendid feast.[43]

Long before then, however, a confusion had arisen regarding the bird's name. Just as with the chicken and the egg, people often wonder whether the bird (turkey) is named after the country (Turkey), or the other way around. The connection between the two is especially puzzling, given that a bird that originated in the Americas shares its name with a nation far to the east, straddling the border between Europe and Asia.[*]

There are two theories as to how the name of the bird arose.[44] It has been suggested that the first turkeys to arrive in Britain were shipped from Spain by merchants from Constantinople (now Istanbul) in Turkey. This is supported by the fact that the British tended at the time to append the word 'Turkey' or 'Turkish' to a whole range of products that originated in that region or beyond.

The second theory is that the British already called the guinea fowl (which originated in Africa), the turkey, for the same reason: that it was shipped to Britain by Turkish merchants. So, when a

---

[*] The recent decision, in June 2022, to renounce the anglicised version 'Turkey' in favour of the name used in the country itself, Türkiye, would have saved a lot of confusion.

TEN BIRDS THAT CHANGED THE WORLD

similar – albeit far larger – bird arrived, it was simply given the same name.

A combination of the two theories is probably correct. Indeed, the name could be even older: Schorger notes that both domesticated peafowl and wild capercaillies (whose males resemble a small turkey cock) were called 'turkeys', perhaps as early as the fourteenth century.[45] This identity crisis is also reflected in the wild turkey's scientific name, *Meleagris gallopavo*, which roughly translates as 'peafowl-cock-like-guineafowl'.[46]

The first published reference to 'turkey' (as a bird) is from a work published in 1555; by the time Shakespeare wrote *Henry IV Part I*, in 1598, the name was in sufficiently wide usage for the Bard to have no need for any explanation.[47]

Whatever the origin of the turkey's name, in a bizarre twist, the Pilgrim Fathers took domestic turkeys on their long sea journey across the Atlantic, bringing them back to the land where their wild ancestors still lived. When these pioneers finally managed to explore their new home in Massachusetts, they must have been astonished to find the counterparts of those same birds, roaming free in the forests.

Hunting was a major source of meat and, being so large and tasty, abundant and (at that time) often fearless of humans, wild turkeys made an ideal quarry. As one early colonist, William Strachey, noted in 1610, turkey meat was 'the best of any kind of flesh which I have ever eaten there'.[48] Settlers often made use of skilled indigenous American hunters to help them track down, kill and catch these birds. One early-seventeenth-century account records that, even after the turkeys had been shot, and lost the power of flight, they still needed to be chased down and caught, otherwise they would run away and hide in the forest.[49]

Other methods by which the birds were hunted included catching them in nets as they fed on fallen acorns, using live decoys, encircling them before driving them into a pen, and firing arrows through a blowgun – a method mainly used, with some success, by children. Turkeys were also easier to hunt in harsh winters, as they tired easily when trying to run through deep snow.[50]

Another way to find wild turkeys was to make a loud noise. This arose from the early discovery that – like other gamebirds, including the peafowl and pheasant – they are quick to respond to any unexpected or unusual sounds. The artist and explorer John James Audubon observed that when a tree was felled in a forest, any turkeys within hearing range would begin to make a loud gobbling call, possibly to alert others to potential danger. Ironically, by doing so, they were then often tracked down and shot.[51]

By the early nineteenth century, turkeys were being kept and raised in such numbers they had become a staple food: in 1833, after attending a Philadelphia dinner party at which turkey was served, Captain Thomas Hamilton wrote of its apparently beneficial effect on his fellow diners' manners: 'No man can say a harsh thing with his mouth full of turkey, and disputants forget their differences in unity of enjoyment.'[52]

It was about this time that the (largely unjustified) notion began to take hold that turkeys are less intelligent than other birds. This may have arisen through the birds' reluctance to run or take flight until cornered, or because when they gathered to roost at night, often in the low branches of trees, they could be easily shot.[53] Whatever the reason, the turkey is still considered proverbially stupid. According to Mark Cocker, who chronicled bird-based cultural beliefs in *Birds and People*, one widespread urban myth suggests that if a turkey looks up towards the sky when it is raining, it will eventually swallow so much water that it will drown.[54] Despite having been thoroughly debunked, this is still widely believed.

Cocker also notes the surprisingly recent origin – as late as 1944 – of the word 'gobbledegook' (or 'gobbledygook'), whose dictionary definition is 'Language or jargon, especially in bureaucratic or official contexts, which is pretentious, long-winded, or specialised to the point of being unintelligible to the general public; nonsense, gibberish.' The word derives from the phonetic representation of the strange, throaty sound made by male turkeys, often represented as 'gobble gobble', first recorded much earlier, in the early eighteenth century.[55]

The word 'turkey' is also used, either on its own or in conjunction with another word or phrase, in a variety of other contexts, few of which appear to have any connection with the bird itself. People – especially politicians – 'talk turkey' when they want to have an honest, frank discussion about a tricky topic.[56] Originating as early as 1921, the phrase 'cold turkey' has long since lost its innocent culinary meaning, and refers to the agony of an addict as they wean themselves off their dependence on drugs or alcohol.[57] And in the theatre and film industry, 'turkey' is widely used to describe a play or movie that it is a box-office or critical disaster, defined by the *Oxford English Dictionary* as 'an inferior or unsuccessful cinematographic or theatrical production, a flop; hence, anything disappointing or of little value'. These very varied usages bear witness to the central place of the turkey in the language and culture of the United States.

Turkeys – especially farmed birds – also have a (deserved) reputation for being aggressive. This derives from the courtship behaviour of wild birds, where the much larger male must fight off his rivals in order to mate with the females. One of the most bizarre examples of aggression from domestic turkeys came during the siege of Khartoum, in Sudan. Writing in his diary in September 1884, just a few months before he was killed, the British Governor-General Charles Gordon wrote that one male bird killed two of its own offspring. Nevertheless, Gordon had a sneaking admiration for the species, writing that 'a Turkey-cock, with every feather on end, and all the colours of the rainbow on his neck, is the picture of physical strength'.[58]

Stupidity and aggression aside, many other observers agree with General Gordon on the bird's positive qualities. Albert Hazen Wright refers to the species by the epithet of 'America's noblest game bird', a phrase first used by Audubon.[59]

Today, however, our mass consumption of turkeys is being called into question, leading many to ask whether our long-standing custom of eating turkey at Thanksgiving and Christmas is sustainable, or indeed desirable.

The world's population is growing and, despite a rising trend towards vegetarianism and veganism, especially in the developed world, so is the demand for meat. Turkey is no exception. In just one year, 2016, the global market for turkey meat rose by 9 per cent, to over 6 million tonnes, and is forecast to reach close to 7 million tonnes a year by 2025.[60]

In 2020, in the US alone, consumers bought almost 2.4 million tonnes of turkey meat. This works out as an incredible 7.5 kilos for every American man, woman and child: almost double the figure from 1970.[61] It's worth noting, however, that they do not actually eat it all: at Thanksgiving alone, an estimated 91,000 tonnes of turkey meat are thrown away – roughly a single generous portion per person.[62]

As you might expect, consumption of turkey meat is highest in developed nations such as the USA and Canada, Europe and parts of Latin America. But rising living standards, and a shift towards Western diets in Asian countries – notably China and India, with their vast populations – is also driving the rising trend. Good news for American turkey farmers, who between them supply almost half the world's turkeys.[63]

But perhaps not for the turkeys themselves. Animal welfare campaigners are becoming increasingly concerned about the methods by which so many millions of turkeys are being farmed. One organisation, Advocates for Animals, itemises 'overcrowding, physical mutilations, the thwarting of natural instincts, rapid growth, poor health and hygiene, and inhumane transport and slaughter practices'.[64] It points to a major transformation in turkey-farming practices since the 1970s. Gone are the small, family-run farms that were once scattered across rural America, to be replaced by vast, industrial hatcheries, each over twenty times the size of their 1975 counterparts. From there, the new-born turkeys are sent to huge barns, each containing up to 10,000 birds in an area of roughly 2,300 square metres. This means that each bird has less than one quarter of a square metre of space.

Just how small this is can be hard to imagine. So consider this: the average floor space of a typical house in England and Wales is 99 square metres. Under the current rules, that would be enough room to house over 400 turkeys. Moreover, to stop these stressed and overcrowded birds mutilating one another, the ends of their beaks and toes are cut off – without anaesthetic – causing pain, distress to the birds, and sometimes early death.

Nor, as the campaigners point out, is the term 'free-range' much of an improvement. The term is essentially meaningless, as the only legal requirement is to give the turkeys 'access' to the outdoors; whether or not they choose to take advantage is doubtful, especially during cold winters.

Perhaps the most shocking statistic, however, is the rate of growth of these unfortunate birds. Modern techniques of genetic manipulation have made turkeys able to develop much faster than could ever be achieved naturally. According to a leading farming newspaper, this has led to almost unimaginable growth rates: 'If a seven-pound [human] baby grew at the same rate that today's turkey grows, when the baby reaches eighteen weeks of age, it would weigh 1,500 pounds [680kg, or 107 stone – roughly the equivalent of eight adult male humans].'[65] The huge (and frankly disproportionate) size and weight of the modern farmed turkey means that today's birds can hardly walk, let alone fly. They are very susceptible to a range of diseases, so are pumped full of antibiotics, and are so fat they even need to be artificially inseminated, using syringes, in order to produce the next generation of chicks.

Domestic turkeys do lead mercifully brief lives: they are usually killed at around three to four months old. The slaughtering process is not for the squeamish: the birds are first transported, sometimes for as long as thirty-six hours without food or water, then shackled, hung upside down and stunned using an electrical current. Following this, their throats are slashed, and they are dipped into a tank full of boiling hot water to loosen the feathers for mechanical plucking. Horrifyingly, the stunning process is not always effective, so some birds are likely to suffer death by slashing or scalding.[66]

You may now not be feeling quite so keen to tuck into that plump, roasted turkey at Thanksgiving or Christmas. But even if you do, you would do well to be wary. Every year, on both sides of the Atlantic, a significant number of diners fall victim to food poisoning, thanks to an undercooked or contaminated turkey.[67] Of the one million cases of food poisoning each year in the UK, a higher proportion occur at Christmas than at any other time. Turkey is often the culprit.[68]

The main problem is that most people are simply not experienced at preparing and cooking such a large bird – often five or six times heavier than a standard chicken. One cardinal sin is not properly defrosting frozen turkeys (a large bird can take up to four days to fully thaw out); another common mistake is washing the turkey – ironically because people assume this makes it safer to eat.[69] Undercooking is a perennial problem, though perhaps less so with modern fan ovens. I can recall my grandmother turning on the stove on Christmas Eve, then getting up at the crack of dawn on Christmas Day to put the bird on to cook; even then, we rarely ate our dinner until mid-afternoon.

Even if all these potential pitfalls are avoided, food poisoning can also occur long after the festive meal itself. In Helen Fielding's bestselling book (and later film) *Bridget Jones's Diary*,[70] Bridget's mother's week-old turkey curry, made by reheating the cold leftovers in a sauce, is a classic way of causing a mass outbreak of food poisoning among family, friends and neighbours. Poor food hygiene doesn't just lead to a few days of sickness and discomfort; it can also kill. In December 2012, a forty-six-year-old woman died after eating a Christmas Day lunch at an east London pub, while thirty-three other diners fell sick.[71] Because Thanksgiving generally occurs over a shorter period than Christmas and New Year, it perhaps presents less of a risk. Even so, people have died from salmonella poisoning caught from eating contaminated turkey, at what should have been a celebration, not a wake.[72]

Such incidents are mercifully rare. But in recent years attention has turned to another aspect of turkey production and consumption: its carbon footprint. This is a relatively recent term – first appearing in print as recently as 1999 – and is defined by the *OED* as 'the environmental impact of a particular individual, community, or organisation, or of a specific event, product, etc., measured in terms of the total associated greenhouse gas emissions.'

The good news for turkey fans is that – at least relative to other meats such as beef, pork and lamb – it has a lower carbon footprint. The bad news is that it still appears in the 'red' category: the energy consumption required to produce the 110gm portion most people would eat in their Christmas dinner is equivalent to the greenhouse gas emissions produced by driving over 4 km in a petrol or diesel car.[73] Typically, a turkey dinner – with all the trimmings – results in more than twice the emissions of its vegan equivalent.[74]

Moreover, as the environmentalist Peter Singer explains, the grain used to feed turkeys could go to feed starving people, producing far lower greenhouse gas emissions. The US physician Michael Greger, meanwhile, has suggested that the link between consuming poultry and cancer may be due to carcinogens in cooked meat. For reasons as yet unknown, these carcinogens appear to build up more in the muscles of chickens and turkeys than in those of other animals. So, if you want to stay healthy – and help save the planet – the message is clear: avoid eating turkey.

Might there be a safety net for those of us who still hanker after the full-on roast turkey experience at Christmas or Thanksgiving? Recent developments in the production of artificial (also known as cultured) meat – which looks and tastes almost exactly like its animal-based equivalent – suggest that this laboratory-produced alternative might be economically viable just a few years from now.[75] Rather like the idea of nuclear fusion producing unlimited, environmentally benign, supplies of energy, cultured meat could – in theory at least – bring the factory farming of animals to an abrupt end. There are, of course, many obstacles to its rapid progress, ranging from potential consumer resistance to the more sinister vested interests of the global agriculture industry. But if it can be made to work on the scale required, the day of the domestic turkey may soon be over.[76]

*

Which brings us back to the wild turkey itself, and its varied fortunes over the last century or so. As early as the 1670s, just half a century after the Pilgrims arrived, observers were already becoming concerned about its rapid decline, with one report stating that a combination of the European settlers and native peoples had 'destroyed the breed, so that 'tis very rare to meet with a wild Turkie in the woods'.[77]

By 1842, the situation was becoming critical: 'The Wild Turkey, which was formerly common throughout our whole country, has everywhere diminished with the advancement of the settlements, and is now becoming exceedingly rare in all parts of New England, and indeed in all the eastern parts of the United States.'[78]

Audubon concurred, noting in *The Birds of America*, published from 1827 to 1838, that the species was becoming 'less plentiful in Georgia and the Carolinas' and 'less numerous in every portion of the United States'; worryingly, he also wrote that 'In the course of my rambles through Long Island, the State of New York, and the country around the Lakes, I did not meet with a single individual.'[79]

By 1920, the wild turkey had already vanished from no fewer than eighteen of the thirty-nine states where it had once been found. This happened through a combination of over-hunting and habitat loss, as woodlands (including oak and chestnut trees, on whose seeds which the turkeys feed) were chopped down and replaced with arable crops.[80]

The decline continued, even though interventions were made to try to boost wild populations, including giving the birds supplementary food in winter; planting seed-bearing crops such as millet, wheat and corn; controlled burning of areas of habitat to encourage the regeneration of vegetation on which they would feed; and rearing wild turkeys in captivity before releasing them. These measures sometimes worked on a local scale, but still could not halt the relentless fall in numbers. By 1973, the US population had dropped to just 1.5 million birds.[81] At that point, the various state and federal conservation agencies realised that something radical had to be done to save the wild turkey from potential extinction. While it may seem absurd that this common, widespread and iconic bird might vanish for ever, the demise of the once abundant passenger pigeon and Eskimo curlew meant it was all too possible that the turkey might go the same way.

In the late 1980s, the authorities finally embarked on a programme of trapping the birds where the relict population had become too small to survive, and transporting them to prime habitats elsewhere. This was well-funded: a total of $412 million was raised and spent on saving this iconic American bird. This and the other measures eventually bore fruit: today, numbers have more than quadrupled, to almost seven million.[82]

The future of the wild turkey, a bird so beloved by Americans that Benjamin Franklin once suggested it should have been chosen ahead of the bald eagle as the symbol of the USA (see chapter 8), appears assured. But that cannot be said of the subject of the next chapter: the proverbial figurehead of the finality of extinction – the dodo.

4

DODO

Raphus cucullatus

The dodo never had a chance. He seems to have been invented for the sole purpose of becoming extinct.

Will Cuppy[1]

The bird was like no other I had ever seen. Its dirty grey plumage looked as if it needed a good wash, while the feathers were not neat and flat, as on most birds, but fur-like and fluffed up, like a much-loved and rather tatty teddy-bear. The wings – if you could even call them that – were ridiculously short and stubby; the tail, equally so. The legs were a faded brownish-yellow shade, the body stout, the posture upright. In my naïve, child's eye view, it looked like a deformed turkey.

But the most striking feature was its head. The bird's face was completely bare of feathers, and a dull yellow colour, with a thin hood of feathers like my grandmother's headscarf. The bill was huge, hooked and black, reminding me of a vulture. The ridiculously tiny eyes stared back at me in a curious blend of accusation and sorrow.

For a moment, I thought the bird was about to step towards me, and I flinched in fear. Then I realised there was a thick plate-glass window between us; and that the bird was not only dead, but had been so for more than 300 years.

I must have been about seven years old. I was with my mother, on our annual visit to London's Natural History Museum – for me, the highlight of the summer holidays. The bird, of course, was a dodo – which, even then, I realised was probably the best-known bird in the world, despite being well and truly extinct.

Now, more than half a century later, I realise that my seven-year-old self had been cruelly deceived. I, and the thousands of visitors who stopped to stare at this incredible creature every day, naturally assumed that we were looking at an authentic dodo: stuffed, mounted and put on display as the key exhibit in the museum's 'Extinct Birds' gallery.

But we were mistaken. Everything I could see was artificial. The head, beak and other bare parts had been moulded from plaster; the feathers were from swans and geese, dyed to produce that unappealing dirty grey colour. The Irish writer Roisin Kiberd perfectly sums up its rather dubious status: 'It's a composite, a Franken-dodo made out of parts of other birds sewn together in the image of something never seen in real life by its creator.'[2]*

Even the shape is wrong. The model, made by the celebrated Victorian taxidermist James Rowland Ward, is considerably fatter than the real dodo is likely to have been. This error was not entirely Ward's fault: he had based his work on an early-seventeenth-century portrait of the bird by the Dutch artist Roelant Savery.[3]

Savery, in turn, was following eyewitness accounts of the dodo from the first men to see it in the flesh: a crew of Dutch and other European sailors who had landed on the Indian Ocean island of

---

* In the summer of 2022, more than half a century after my first childhood encounter with the dodo in London's Natural History Museum, I paid a return visit. The bird is still sitting in its glass-fronted cabinet, still looking like a cross between an oversized chicken and a tatty teddy-bear, and still staring balefully back at me and the hordes of visiting children. One thing has changed, though. The museum authorities have come clean about the whole fakery business, with a caption telling us that this is not in fact a real dodo, but a 'Model/Life-sized reconstruction'.

Mauritius in 1598. It has also been suggested that those dodos brought to Europe were overfed on the long sea voyage; this, combined with a lack of exercise, led them to put on weight.[4]

In an artistic version of Chinese whispers, the 'fact' that dodos were ridiculously obese has been passed on down the centuries. It has helped to create the distorted image of the dodo that so imprinted itself on my young mind, which continues to be the default representation of the species from the time it was discovered to the present day.

This distortion matters, because the dodo has become the definitive emblem of extinction: a bird whose very name conjures up the terrible finality of its fate, as in the phrase 'as dead as a dodo'.[5] If we think of it as fat and stupid, that makes it easier to believe that the dodo was in some way complicit in its own demise. The truth – that we, and we alone, are to blame – is rather less comforting.

The apparent stupidity of the dodo has also been linked with its name. The *Oxford English Dictionary* suggests that this derives from a Portuguese word *doudo*, meaning simpleton or fool,[6] though the historian of extinction Errol Fuller has proposed a different, Dutch origin, from the plural *dodaersen*, which he translates as 'fat behinds'.[7] Neither is very flattering.

In English, 'dodo' has a comical ring to it which, when linked to the belief that the birds were both obese and unintelligent, and the fact that they were unable to fly, again suggests that they were to blame for their own fate.

The fundamental contradiction about the dodo, as Fuller points out in his wonderfully detailed monograph of the species, is that, for all its fame, the dodo represents a paradox: 'despite its enormous popularity and the great proliferation of dodo literature, we know almost nothing about the bird itself.'[8] Indeed, it has be argued that we know more about the biology of Tyrannosaurus rex – the celebrated dinosaur that went extinct some 66 million years ago – than we do about the dodo.

This is the paradox of one of the most extraordinary stories in the long and eventful history of humanity's relationship with birds. It demonstrates what can happen when, with hardly any tangible evidence to go on, we create false – yet oddly enduring – myths about the past.

The dodo's disappearance heralded what has become known as 'the Age of Extinction'.[9] In the 300 years or so since it died out, we have lost hundreds more species, some so obscure they are known from just a single museum specimen, others – such as the great auk and passenger pigeon – almost as famous as the dodo. Recently, the pace of extinction has sped up exponentially, with roughly one in seven of the world's 10,700 species of bird now considered at risk. Their plight, which echoes that of the dodo, reveals some harsh truths about our attitudes towards the wild creatures with which we share our planet.

Since the dodo was first discovered, just over 400 years ago, it has been celebrated, maligned, and ultimately turned into a caricature of itself. It has become the definitive symbol of loss, and has shown up both the blinkered nature of organised religion and the legacy of bad science. It has led to dispute, rivalry, profiteering, fraud and skulduggery. It is, in many ways, the ultimate triumph of myth over reality.

But we should not forget that, at the heart of this story, is a single species of bird, which only now is starting to emerge from this muddled melee of misrepresentation, parody and distortion. A bird which, by its own rapid and unfortunate demise, changed the way we look at the natural world for ever.

Let's start with the little we actually know about the dodo. From the start, accounts of the bird were mired in confusion, with a range of often incompatible descriptions of its size, shape, colour and overall appearance.[10]

The earliest record comes second-hand, from an English translation of a now lost Dutch original. Published in 1598, it was written less than a decade after European sailors first landed on Mauritius, where they found a 'great quantity of foules twice as big as swans . . . [and] finding an abundance of pigeons and poppinayes, they disdained any more to eat of those great foules calling them Wallowbirdes, that is to say loathsome.'[11]

It is sad, but in the light of later events predictable, that not only does this first account contain a blatant exaggeration of the dodo's actual size, but it also focuses purely on whether or not the bird was edible. Later accounts from the same voyage did reduce the dodo's size to 'as big as a swan', but continued to focus on the understandable preoccupation of long-distance voyagers: the ready availability of fresh meat.

Gradually, though, these contemporary accounts begin to add inviting snippets of detail about these strange and unfamiliar birds. We learn that they had 'a round rump with two or three curled feathers', 'no wings but instead have three or four black quills', 'the body of an ostrich . . . and on the head a veil as though it were wearing a hood' and, perhaps most strikingly, that 'they walked upright on their feet as though they were a human being.'[12] A later writer describes the dodo as: 'more especially distinguished from other Birds by the Membranous Hood on his Head, the greatness and strength of his Bill, the littleness of his Wings, his bunchy Tail, and the shortness of his Legs'.[13]

The dodo's structure and appearance were so bizarre that, for more than two centuries after its discovery, scientists remained baffled as to its relationship to other birds. At various times it was considered to be related to the ostriches, to rails, albatrosses, even vultures (perhaps suggested by that hooked bill).[14]

When, in 1842, the young Danish zoologist Johannes Theodor Reinhardt tentatively suggested that dodos might actually be related to pigeons, an observation based on his careful examination of a museum skull,[15] he was widely ridiculed. Six years later, though, Reinhardt was vindicated, when two British naturalists, Hugh Edwin Strickland and Alexander Gordon Melville, published the first authoritative monograph of the species.[16] Having examined the morphology and anatomy of a mummified dodo head kept at the Oxford University Museum of Natural History, and a dried foot housed at the British Museum in London, they too concluded – correctly, as it turned out – that the flightless dodo, and its now also extinct cousin the Rodrigues solitaire, were related to pigeons and doves.

Their verdict has stood the test of time. The dodo and solitaire are indeed members of the family *Columbidae*.[17] This is one of the largest of all bird families, with almost 350 species found across six of the world's seven continents. But with no fewer than twenty-one species (and a further eleven subspecies) having already gone extinct in historical times, it is also one of the most vulnerable.[18]

The dodo's closest living relative is the Nicobar pigeon, found across a broad swathe of South-East Asia from the Andaman and Nicobar Islands, across the Malay Peninsula, the Philippines and Indonesia, to the Solomon Islands. However, the much rarer tooth-billed pigeon (confined to the Pacific archipelago of Samoa) is far more similar in appearance, sporting a large and powerful bill, rather like a smaller version of the dodo's. This species – sometimes known as the 'little dodo' (from its scientific name *Didunculus*) – is now, sadly, also on the verge of extinction.[19]

Recent studies have found that the ancestors of the dodo and solitaire split from the rest of the pigeon family as long as 23 million years ago. We know that they must have still been capable of flight, simply because the remote islands where they ended up did not emerge from the Indian Ocean until between 8 and 10 million years ago, as the result of undersea volcanic activity.[20]

Over many millions of years, along with other birds confined to oceanic islands, the dodo's ancestors gradually lost their ability to fly, which in this predator-free environment entailed a needless expense of energy. There were also no herbivorous mammals competing for food in their new home so, for much of their existence, they thrived. Then, one day in 1598, a Dutch expedition made landfall on Mauritius, and the dodo's fate was sealed.

It is often suggested that the dodos were wiped out because they were killed for food. But the original Dutch name for the species – *Walghvoghel*, meaning 'tasteless bird' – suggests that the sailors soon got fed up with the dodo's flesh, preferring other, smaller and tastier species such as pigeons. The truth, it seems, is rather more prosaic: it was the introduction of domestic animals such as pigs, ship's stowaways including rats, and exotic pets such as crab-eating macaques, that really did for the dodo. Being a ground-nesting bird meant its eggs and chicks were especially vulnerable to being eaten by these new arrivals.

Whatever the reasons for its rapid decline, barely six decades after the Europeans arrived on Mauritius, the dodo was, if not yet extinct, too far gone to recover. Even today, though, there is some dispute over when the species finally vanished. The most widely accepted date is 1662, when the shipwrecked Dutch sailor Volkert Evertsz captured what may have been the last of the species, on a small island just off the main one:

> We drove them together into one place in such a manner that we could catch them with our hands, and when we held one of them by its leg, and that upon this it made a great noise, the others all on a sudden came running as fast as they could to its assistance, and by which they were caught and made prisoners also.[21]

To confuse the issue, however, there are a number of reports of dodos encountered some years after 1662, though it has been suggested that the birds in question may actually have been the red rail (also known as the Mauritius red hen), another flightless endemic bird which went extinct soon after the dodo.[*]

In one sense, the exact date hardly matters. What is more important is what happened next: how the dodo rapidly turned from a real, flesh-and-blood bird into an icon – indeed, *the* icon – of extinction. As the US humourist Will Cuppy noted, 'The dodo never had a chance. He seems to have been invented for the sole purpose of becoming extinct.'

Yet the dodo's story is more complex and ambivalent than it might first appear, as the author and naturalist Michael Blencowe explains:

> The dodo is an enigma. To be as dead as one is the final word in finality. Yet the dodo has transcended death, becoming caricatured, objectified, merchandised and resurrected as an unlikely and somewhat dumpy messiah. The dodo is one of the most recognisable species that has ever lived. It has achieved a dubious immortality: the smiling face of extinction.[22]

What can we learn, then, from the dodo's physical and cultural destiny through the ages, and why is this so relevant today?

*

[*] The theory is that sailors visiting the island in the late seventeenth century were naturally expecting to see dodos. So, when they came across another large, flightless bird – the red rail – they may have assumed this was the species they were looking for; whereas by then, in all likelihood the dodo was already extinct. A. S. Cheke and J. C. Parish, 'The Dodo and the Red Hen, A Saga of Extinction, Misunderstanding, and Name Transfer: A Review', *Quaternary*, 3(1): 4 (2020).

Even before the dodo went extinct, it was already being treated as a kind of avian circus act. During the brief period when the species lived alongside us, the handful of dodos that survived the long and arduous sea voyage back to Europe from Mauritius were put on show alongside other 'freaks of nature', to titillate and entertain the paying public. We know this because, sometime around the year 1638, the English writer and historian Hamon L'Estrange was strolling with his companions through the streets of London, when he noticed 'a picture of a strange looking fowle hung out upon a clothe' outside a property. Curious, he entered the premises, where he saw 'A great fowle somewhat bigger than the largest Turky cock, and so legged and footed, but stouter and thicker and of a more erect shape, coloured before like the breast of a young cock fesan [pheasant], and on the back a dun or dearc [dark] colour. The keeper called it a Dodo.' L'Estrange questioned the dodo's keeper about a heap of pebbles on the ground nearby, and was told that the bird would eat them to aid its digestion. Having marvelled for a while at this bizarre creature, he and his companions took their leave.

This frustratingly brief account proves that at least one dodo – and probably others – were taken into captivity, brought to Europe, and managed to survive long enough to be put on display.[23] Errol Fuller speculates on a very different fate for the species if it had managed to survive just a little longer. Had it been able to establish itself in our city parks and gardens, he wonders if 'today, dodos might be as common as peacocks in ornamental gardens the world over!'[24] Instead, 'all that remains are a few bones and pieces of skin', together with paintings and written descriptions which 'are curiously inadequate in the information they convey'.[25]

To properly understand the dodo's cultural journey, from a living bird to the emblem of extinction, we must consider the religious and philosophical climate at the time of its disappearance, and especially the way people viewed the creation of species.

The fundamentals of Christian belief – that an all-powerful God created all living things – are stated clearly at the start of the Old Testament, in the first chapter of the Book of Genesis:

> And God created great whales, and every living creature that moveth, which the waters brought forth abundantly, after their kind, and every winged fowl after his kind: and God saw that it was good.[26]

What is hard for the modern reader to comprehend is that if an all-powerful Creator had given life to 'every winged fowl' (and presumably flightless ones, too), then the notion that He could then allow any species to go extinct was nothing less than heresy.

The prevailing orthodoxy, developed by the medieval church but still holding sway several hundred years later, was known as 'the Great Chain of Being'. This proposed a hierarchical pyramid, with God at the top, below whom came the angels, then human beings, followed by animals and plants.[27] The key belief that held the Great Chain of Being together was that, having been created by God, it was perfect, unbroken and (most importantly) unchanging. Therefore, extinction of any species was fundamentally impossible, as God could not allow any links in the chain to be broken and disappear.[28]

So even though it has frequently been suggested that the dodo was the first creature to alert the world to the impending ecological crisis of man-made extinction, the story is a lot more complex – and far messier. For more than a century after the dodo had disappeared, that a species could simply vanish was simply not contemplated. Far from the dodo's disappearance being the 'light bulb moment' that alerted the world to the huge and terrible peril of permanent extinction, it passed almost unnoticed. Only at the end of the eighteenth century, more than a hundred years after the last dodo was seen, did the French zoologist Georges Cuvier first propose that a species could indeed go extinct. He did so by demonstrating that wild creatures that had once roamed the Earth, such as mammoths and mastodons, were not – as some had suggested – still living in remote regions of Africa, but had died out for ever.[29]

This marked a key moment in the broader movement known as the Enlightenment. European philosophers, scientists and intellectuals were beginning to question – and ultimately abandon – the outdated, yet still hugely influential, religious beliefs of the Church, and replace them with ideas based on field observation and logical deduction. This would culminate in the most important scientific breakthrough of all: Charles Darwin and Alfred Russel Wallace's Theory of Evolution by Natural Selection (see chapter 5).

It is important to note that although Cuvier clearly believed that species could go – and had gone – extinct, he did not then proceed to the next logical step: that humans might, either directly or indirectly, be to blame. To be fair, this was because he was focusing mostly on fossil remains, many of which pre-dated human life on Earth.

Meanwhile, the Church was still unable to entertain the concept of extinction at all, as its later reaction to Darwin and Wallace's revolutionary theory comprehensively demonstrated. Before Cuvier, as the science writer Colin Barras points out, the very idea that a species might go extinct was simply unthinkable; consequently, the specimens of dodos that had reached Europe were not accorded the importance they should have been.

Because scientists were still unable to accept that a species could disappear for ever, museum curators took a cavalier attitude towards the dodos in their possession: after all, if one became lost or damaged, they could always obtain a replacement. The result was that by the early decades of the nineteenth century, not a single complete skeleton of a dodo had survived.[30] This explains the fake 'dodo' in the Natural History Museum that so enthralled me as a child. As Barras concludes, 'The dodo was lost for a second time.'[31] Until the second half of the nineteenth century, the sum total of physical specimens of the dodo would number just three, all of which are listed and shown in Errol Fuller's 2002 monograph on the species:[32] the remains of a head and foot kept at the Oxford Museum of Natural History, which are the only remains of a dodo's soft tissue anywhere in the world;[33] a skull at the Natural History Museum in Copenhagen, Denmark;[34] and an upper jaw at the National Museum in Prague, Czech Republic.[35] In addition, the foot of a dodo, originally housed at the British Museum, and then at the Natural History Museum, was subsequently lost.

Then, in 1865, came the most important breakthrough in our knowledge and understanding of dodo taxonomy. A schoolteacher on Mauritius, George Clark, heard of the discovery of a number of bones in a local swamp, the Mare aux Songes.[36]

Clark had emigrated from his native Somerset to Mauritius almost thirty years earlier, to teach at a local religious school. He was a keen naturalist and fossil hunter, and having discovered, to his great surprise, that there were no remains of dodos on the island itself, he determined to spend his spare time searching for them. But since then, almost three decades had gone by, with absolutely no success.

Then came the news he had been hoping and praying for. While supervising excavations to build a railway line across the swamp, a young engineer named Harry Higginson had discovered a cache of remains. On examining them, Clark realised that they had finally stumbled across the ornithological equivalent of the Holy Grail: several hundred bones of the dodo.

Clark later sold most of these to museums and private collectors in the United Kingdom, sending one consignment to the celebrated biologist and palaeontologist Richard Owen.[37] He also sent a beautifully preserved set of specimens to his friend Dr William Curtis, founder of the eponymous Curtis Museum, in the Hampshire town of Alton.[38]

By coincidence, Higginson and Clark's discovery in 1865 coincided with the publication, that very same year, of the children's book *Alice's Adventures in Wonderland*. Written by Lewis Carroll, the pen name of the eccentric Oxford don and mathematician Charles Dodgson, this curious tale features a bizarre cast of characters, including the Mad Hatter, Mock Turtle, White Rabbit and Dodo. In one of the book's most memorable scenes, the dodo proposes a race, in which there are no set rules, and so, in a phrase that has since become proverbial, 'Everybody has won, and all must have prizes.'[39]

The character of the dodo, superbly depicted by the renowned artist John Tenniel, soon captured the public imagination, and helped cement the image of the bird in popular culture. But of all the creatures he could have chosen, why did Dodgson pick the dodo in the first place? One widely held (but sadly unproven) theory is that he was making fun of himself; more specifically, his stammer, which at moments of stress would apparently cause him to introduce himself as 'Do-Do-Dodgson'. Yet he would also have been well aware of the dodo's role as an archetype of extinction. This idea had first been aired in a popular publication aimed at educating the working classes, the *Penny Magazine*, in 1833, a year after Dodgson's birth. The author, naturalist William Broderip, presciently wrote:

> The agency of man, in limiting the increase of the inferior animals, and in extirpating certain races, was perhaps never more strikingly exemplified than in the case of the Dodo. That a species so remarkable in its character should become extinct . . . is a circumstance which may well excite our surprise, and lead us to a consideration of similar changes which are still going on from the same cause.[40]

It is interesting to note that, almost two centuries ago, Broderip was proposing the idea that extinction is not some random and trivial event, but has a wider global significance. In 1848 Strickland and Melville's *The Dodo and its Kindred* was published, a book which has been described as kick-starting 'dodo-mania', even though, as has been pointed out,[41] Strickland and Melville, while acknowledging the part played by 'human agency' in the dodo's demise, nevertheless revert to the tired old idea of human superiority: that 'Man is destined by his Creator to "be fruitful and multiply and replenish the Earth and subdue it".'[42] Though it was a later extinction, that of the great auk in the mid-nineteenth century, that really launched the movement to try to prevent other species meeting the same fate,[43] by 1869 a review of Strickland and Melville's book in *Blackwood's Edinburgh Magazine* was supplanting the great auk with the dodo as the global icon of extinction.[44] 'The death of the Dodo and its kindred is a more affecting fact,' noted the anonymous *Blackwood's* reviewer, 'as involving the extinction of an entire race, root and branch, and proving that death is a law of the species, as well as of the individuals that compose it.'

Today, the dodo is an invaluable flagship species to warn humanity of the current fragility of global ecosystems, along with the growing vulnerability of so many other species being driven extinct by human actions, such as the Yangtze River dolphin of China, and the ivory-billed woodpecker of the south-eastern United States and Cuba.[45]

The dodo may be by far the most famous of all lost species, but it was by no means the only one endemic to the Mascarene Islands (Mauritius, Réunion and Rodrigues) to have gone extinct. No fewer than forty-eight species of these islands' terrestrial vertebrates – mostly birds and reptiles – are known to have disappeared between 1500 and 1800.[46] As well as the dodo, Rodrigues solitaire, and Mauritius red hen, other avian casualties include the broad-billed (or raven) parrot, the Mascarene parrot and the Mascarene coot. All, like the dodo, were driven to extinction by the arrival of European explorers and their destructive menagerie of domestic animals. Ironically, these islands were amongst the most recent places on the planet to be colonised, and yet are amongst the most ecologically damaged; 'thus illustrating', as one conservationist dryly notes, 'our great propensity to destroy the environment'.[47]

The Mascarene Islands may have seen a more rapid loss than many other places, but they were far from unique in losing such a high proportion of their bird life. Recent studies have shown that birds found on oceanic islands have a forty times greater chance of being threatened by extinction than those living on larger, continental landmasses. As a result, of all the world's threatened bird species, 39 per cent are confined to islands.[48]

There are several reasons why this should be so. As the story of the dodo demonstrates, island species have often adapted to the very specific conditions there – notably a lack of predators – by becoming flightless, or near-flightless. This makes perfect sense – until human beings enter the picture, when flightlessness rapidly turns into a disaster. Birds that cannot fly are, as we saw with the dodo, very easy to hunt and kill, and easy prey for any invasive, non-native species. As in all but the most remote and pristine locations around the globe, the loss and degradation of habitat is also a major issue, while a species confined to a single island, or island group, will inevitably have a relatively low population compared with a species living on a larger landmass. That population is also likely to have a lower genetic diversity.

Of all these threats, by far the greatest problem is the presence of invasive and predatory species. A recent BirdLife International report revealed that three-quarters of threatened birds on oceanic islands are at risk from deliberate or accidental introductions – a total of almost 500 in all, comprising nearly one-third of all threatened bird species. Of those introduced predators, rats and cats are by far the most important threats.[49]

This is not a new problem, as the sad tale of the Stephen's Island wren shows. Stephen's Island is a tiny speck of land, covering an area of just 1.5 square km, off the northern tip of New Zealand's South Island. The island was uninhabited until the late nineteenth century, when a lighthouse was built and, in 1894, the keeper – David Lyall – moved in. Unfortunately, Lyall brought with him several cats, one of which was pregnant. Over the course of the year these escaped and went feral, feeding on the various small birds found on the island. One was a small, flightless songbird, later named Stephen's Island wren (also known, ironically given the identity of its unwitting nemesis, as Lyall's wren).

During that spring and summer, the island cats regularly brought the bedraggled corpses of the wren to David Lyall, who sold them to the naturalist and bird collector Henry H. Travers. Travers in turn made a handsome profit by selling several specimens to the wealthy bird collector Walter Rothschild.

The following February, Travers and his assistants visited the island, keen to procure more specimens of the wren. They did not find a single one. One month later, on 16 March 1895, an editorial appeared in the Christchurch newspaper *The Press*, opining that 'There is very good reason to believe that the bird is no longer to be found on the island, and, as it is not known to exist anywhere else, it has apparently become quite extinct. This is probably a record performance in the way of extermination.'

Bizarrely, a plan was then made to eradicate the feral cats from Stephen's Island. Thirty years later, in 1925, this project was finally completed – far too late, of course, for the eponymous wren.[50]

In an even greater irony, it has since been discovered that before the arrival of the Māori people, who travelled across the ocean from eastern Polynesia during the mid-fourteenth century, the Stephen's Island wren was found across much of New Zealand. Just as European explorers inadvertently introduced dogs, cats and rats to Mauritius, so some 250 years earlier, the Māori had brought their own domestic animals to their new home. In the absence of large mammals, the settlers also hunted any large birds they came across for food.

This led to what is arguably the most rapid and destructive cull of endemic species in human history, and the permanent impoverishment of New Zealand's unique avifauna. The victims included no fewer than nine different species of moa, one of which, the South Island giant moa, reached a height of 3.7 metres, making it the tallest bird ever known. The moas' only predator, Haast's eagle, one of the largest and heaviest birds of prey ever to have existed, was also wiped out.

Overall, since the Māori first settled in New Zealand, almost half the country's endemic terrestrial bird species have gone extinct.[51] So much for the widely held belief that these early settlers cared more for the environment than the later, European colonists. As one biologist pithily observes, 'We like to think of indigenous people as living in harmony with nature. But this is rarely the case. Humans everywhere will take what they need to survive. That's how it works.'[52]

It is tempting to suppose, given the number of island species now under imminent threat of extinction, that many more will follow the moas, Stephen's Island wren and dodo into oblivion over the coming decades.

But now, at what surely is the eleventh hour, far-sighted zoologists and conservationists are fighting back, with some of the most ambitious projects ever undertaken. The aim is to eradicate non-native pest species from offshore islands, to enable the species that breed there to recover. Already, several small islands around the United Kingdom, such as Lundy off the coast of Devon, have successfully completed schemes to eradicate non-native brown rats. The results have been almost immediate. Within just fifteen years, Lundy's seabird populations have increased dramatically, with the emblematic puffins (after which the island is named) rising from just thirteen birds to 375, and Manx shearwaters going from fewer than 300 pairs to more than 5,500.[53] Likewise, on the connected islands of St Agnes and Gugh on the Isles of Scilly, not only breeding seabirds but a small mammal, the Scilly shrew, have benefitted from the removal of rats.[54]

The Royal Society for the Protection of Birds (RSPB), which was involved with the Lundy rat eradication project, has recently embarked on a far more ambitious plan: to remove house mice from Gough Island, a British Overseas Territory in the South Atlantic Ocean.[55] The seabird under threat from the mice is, perhaps surprisingly, one of the world's largest: the Critically Endangered Tristan albatross.

In a scenario from a horror movie, the albatross chicks – and sometimes even the adults – are being eaten alive by the mice as they sit on their nest. These rodents, accidentally brought to Gough by sailors in the nineteenth century, have mutated to grow to several times their usual size, and even though their prey outweighs them by as much as 300 times, they are still able to inflict horrific injuries, which cause their unfortunate victims to bleed slowly to death. The mice also eat the eggs of the albatrosses and other seabirds, with an estimated death toll of up to two million birds each year. The original plan was for the eradication team to land on the island in May 2020, before the Covid-19 pandemic led to a year's delay, but now the team is in place and the programme has finally begun.[56]

If this sounds ambitious, consider the latest plan from the government of New Zealand. Predator Free 2050 aims to do exactly that: rid the entire nation of introduced, non-native predators by the year 2050 – less than thirty years away.[57] The first stage, to be completed by 2025, will be to wipe out predators such as stoats, rats and possums from all the nation's island nature reserves, which are home to the vast majority of threatened birds including the kakapo (a flightless parrot), and several species of New Zealand's national bird, the kiwi.[58]

If Predator Free 2050 does succeed, then, using the lessons learned, conservationists might ultimately be able to eradicate non-native predators from all oceanic islands around the globe. That could, even this late in the day, save many hundreds of bird species currently under threat of going the same way as the dodo.

Some island birds have already been saved, against seemingly overwhelming odds. Back on Mauritius, one pioneering conservationist, Professor Carl Jones, has spent more than four decades – and devoted his entire professional life to – saving the island's remaining threatened birds, for which in 2004 he was deservedly awarded the MBE.

Born in Wales, Jones arrived on Mauritius in his mid-twenties in 1979, and immediately began work on a project to save two of the world's most endangered birds, the pink pigeon and Mauritius kestrel, both endemic to the island, and at the time teetering on the edge of extinction.

It must have seemed like a hopeless cause: with just four known individuals alive in the mid-1970s, the kestrel was surely doomed. Even if these birds could be saved and bred in captivity, the thinking went, the gene pool would be so limited that the species could not possibly survive again in the wild.

Well-respected and influential environmentalists such as Norman Myers opposed the plan altogether, on the grounds that the limited funds earmarked to save the species could be better spent elsewhere to try to save the many other threatened (but more populous) bird species.[59]

Perhaps Carl Jones didn't hear what the sceptics said or, if he did, maybe he didn't care. Instead, he threw away the captive breeding rule book, and began devising his own plan, using simple, but ultimately very effective, strategies. These included 'double-clutching', where the first clutch of eggs laid was immediately removed before incubation could start. The pair of birds would then lay a second clutch, while the first one could be artificially incubated and the chicks reared in captivity. Jones also used foster-parents, gave the birds dietary supplements, and eventually began to return the birds raised in captivity to the wild.

In a single decade from 1983 to 1993, the project reared no fewer than 333 Mauritius kestrels, relocating many of these back into their original home, and eventually creating a self-sustaining population.[60] What had once been the rarest bird in the world is now thriving. In a 1994 paper outlining the success of the project, Jones and his co-authors gave pride of place to Myers' opposition to their plans, restraining their understandable urge to add, 'I told you so!'[61]

Jones and his team have also brought other critically endangered species back from the very brink of oblivion. These include the aforementioned pink pigeon, whose numbers have risen from just ten remaining birds in 1991 to almost 500 today,[62] and the island's only remaining parrot species, the echo parakeet, which has also rapidly increased, from ten birds to over 800.[63]

Early on, Jones realised that restoring these rare birds to the wild would only work if suitable habitat existed for them. To ensure this, he and his colleagues have worked tirelessly to remove invasive species, restore ecosystems and enable keystone species to thrive.[64]

In 2016, Carl Jones, a man once described as 'a raving optimist',[65] won the prestigious Indianapolis Prize, known as the 'Oscars' of conservation.[66] 'Without Carl Jones', his employers rightly boasted, 'the natural world would be a much poorer place.' That is, if anything, a major understatement. Thanks to him, and his vision and skills, the Mauritius kestrel, pink pigeon and echo parakeet are still living, breathing species, and have not, like so many island birds, gone the way of the dodo.

On Mauritius, the island where it all began, just how relevant is the dodo today? Judging by its almost ubiquitous presence, very much so. The dodo has been appropriated as an omnipresent symbol, a combination of a 'jolly national mascot' and universal salesman, in the words of the US writer Meg Charlton. 'The dodo lends its name to pizza parlours and coffee shops, its likeness to beach towels and backpacks. There are giant dodo statues in public parks and mall food courts. Countless tourist shops hawk tiny carved dodos for a few dollars.'[67]

Not everyone welcomes the ubiquity of images of the dodo on Mauritius. In a children's book of the 1980s, the first-person narrator, a dodo, laments the seemingly endless dodo-related products on the market: 'What has become of us? . . . Stamps, keyrings, matchboxes, T-shirts, tea towels, pictures made from sugarcane straws or string, bottle openers, stickers, cards, signs, bookends . . . and even a mascot flying in a helicopter, looking more like a blimp than the noble birds we once were.'[68] If a product can be imprinted with a dodo logo, however distasteful, it seems it will be. You can even buy a packet of chocolate raisins, labelled 'Dodo Whoopsies', ostensibly faithful representations of the bird's droppings.[69]

Official initiatives have seen the dodo take pride of place on Mauritian currency, postage stamps – including several sets designed by the dodo expert Julian Hume[70] – and the country's coat of arms. It has become, bizarrely, a symbol of national identity, even though the bird's story is for human beings one of shame and loss. In a supreme irony, the image of a dodo even appears on the immigration form filled in by foreign travellers: the descendants of those whose arrival here caused the dodo to go extinct in the first place.[*]

Meanwhile, archaeologists and naturalists are following in the footsteps of George Clark, searching the Mare aux Songes for more dodo remains, and are still finding them: expeditions from 2005–9 unearthed more than 700 bones, including those of many adults and a chick. This has led to fantasies of a *Jurassic Park* scenario in which DNA extracted from these bones could in theory be used to reconstruct the entire genetic sequence of this lost species and bring it back to life. Who knows, if this were to come to pass, then Errol Fuller's fantasy of dodos wandering around our public parks like peacocks might one day come true.

In *The Song of the Dodo*, his masterful study of the concept and reality of Island Biogeography, the US science and nature writer David Quammen imagines the fate of the very last dodo to walk the Earth:

---

[*] Mauritius, unusually for a European colony, was not inhabited before the Dutch settlers arrived; and so, as Charlton puts it, 'Mauritians are all the descendants of immigrants'. The caption to a cartoon of a dodo on display in the capital's Natural History Museum confirms this. It reads: '*Les vraies Mauriciens ont été mangés par les Hollandais il y a longtemps*' – which translates as 'The real Mauritians were eaten by the Dutch a long time ago.'

In the dark of an early morning in 1667, say, during a rain-storm, she took cover beneath a cold stone ledge at the base of one of the Black River cliffs. She drew her head down against her body, fluffed her feathers for warmth, squinted in patient misery. She waited. She didn't know it, nor did anyone else, but she was the only dodo on Earth. When the storm passed, she never opened her eyes. This is extinction.[71]

In a world where we are finally beginning to recognise – and in some cases accept – the immense destruction wrought by human beings on the rest of the world's wildlife, how can we come to terms with the concept of extinction? It is, like our own deaths, something we simultaneously recognise as inevitable and yet at the same time reject; perhaps because, like our own mortality, it is far too painful to contemplate.

The Australian ethnographer Deborah Rose has called extinction a 'double death' – an elimination of both the future and the past.[72] And as the cultural historian Anna Guasco points out, the dodo is 'the canary in the coal mine of anthropogenic destruction. Its extinction is seen as the inevitable outcome of human interaction with nature.'[73]

It is clearly too late to save the dodo, and too late to save the dozens – perhaps hundreds – of species of bird that have already gone extinct in historical times. It is, sadly, almost certainly too late to save many of those on the brink, such as the regent honeyeater of south-east Australia, whose numbers have become so depleted that its offspring can no longer learn the species' distinctive song from their fathers[74] (see chapter 10). These have become what Australian birder and conservationist Sean Dooley has dubbed 'zombie birds' – still (just about) surviving, but doomed to disappear during the next few decades.[75] It seems we have failed to learn any lessons from the rapid demise of the dodo, as Douglas Adams and Mark Carwardine have pointed out: 'It's easy to think that as a result of the extinction of the dodo, we are now sadder and wiser, but there's a lot of evidence to suggest that we are merely sadder and better informed.'[76]

Meanwhile, in the next chapter, I shall examine what is arguably the most crucial tipping point between religion and science. The story takes place on a remote island group in the equatorial Pacific Ocean and involves not just one species but a baker's dozen: the thirteen or more different birds collectively known as 'Darwin's finches'.

5

DARWIN'S FINCHES

Geospizinae sp.

In the thirteen species of ground-finches, a nearly perfect gradation
may be traced, from a beak extraordinarily thick, to one so fine,
that it may be compared to that of a warbler.

Charles Darwin, *The Voyage of the Beagle* (1839)[1]

The story – or rather the prevailing myth – of Charles Darwin and
his eponymous finches goes like this:

*It had been a long and arduous voyage. The expedition, which had
left the Devonshire port of Plymouth in December 1831, was expected
to return home two years later, but was already well into its fourth
year. Finally, on 15 September 1835, HMS* Beagle *made landfall on the
Galápagos Islands, in the eastern Pacific Ocean.*

*For the ship's naturalist, Charles Darwin, this was the most exciting
moment in his distinguished career. Wasting no time, he immediately
went ashore, and was astounded at what he saw. As well as penguins,
albatrosses and marine iguanas – all ridiculously tame and approach-
able – Darwin's expert eye soon noticed some curious small birds,
which he immediately recognised as finches.*

*Over the course of the next few weeks, on each island that he visited,
he saw similar birds. But, depending on the habitat they lived in, each
showed noticeable differences, especially with regard to the size and
shape of their bills.*

*Darwin had long been searching for proof that his theory of evolu-
tion by natural selection – that different species evolved to suit their
specific environments – was correct. And now, in the dozen or so birds*

he named Darwin's finches, he could see exactly how this had happened.

All of them descended, he realised, from a single common ancestor: a species of the finch family that had somehow reached the islands from mainland South America, over a thousand kilometres to the east. Having arrived on the islands, this one species had, over millions of years, diverged into the dozen or more he could see today.

For Darwin, this was truly a 'eureka moment' – up there with Copernicus's realisation that the Earth went around the Sun, or Sir Isaac Newton's discovery of gravity. Once the Beagle *returned home, a year later, he would be able to write up his theories in his masterwork* On the Origin of Species, *with the finches at the heart of his argument.*

That book would go on to change the world – all thanks to those rather dull-looking, but scientifically exceptional, birds.

It is a fascinating story, which has lost little of its power for being repeated, *ad infinitum*, in endless books, newspaper and magazine articles and websites, and on radio and television programmes.[2] There's just one problem: it is pure myth.

To start with, Charles Darwin was not even the official naturalist on HMS *Beagle*. Rather, as the evolutionary biologist Stephen Jay Gould points out, Darwin was there as a 'gentleman companion' for the ship's captain, Robert Fitzroy, who, by virtue of his social status and position, was not permitted to mix socially with the rest of his officers and crew.[3] Fitzroy also suffered from depression, which it was thought the presence of Darwin might help to alleviate.[*]

---

[*] Sadly, it did not: many years later, on 30 April 1865, Fitzroy took his own life by cutting his throat with a razor.

Just twenty-two years old when the *Beagle* embarked on its famous voyage (Fitzroy himself was only a year older), Darwin was a naïve, callow and rather unfocused young man. He had graduated from Cambridge, and had a lifelong passion for the natural world, and his family and friends hoped that he would gain valuable knowledge and experience of the wider world before returning home with the maturity required for his intended career in the Church.[4]

The voyage, however, turned out to be rather less enjoyable than Darwin had expected, not least because he suffered from appalling seasickness. He also frequently clashed with Fitzroy, who, as a dogmatic evangelical Christian and High Tory, was about as far removed from Darwin's own more liberal beliefs as any man could be. No wonder that, each time the ship landed, Darwin lost no time in mounting an expedition, which (as was usual at the time) involved killing, and collecting, as many wild creatures as possible.

Neither was Darwin close to developing a coherent philosophy, let alone a finished theory, of how evolution actually took place. He would not publish *On the Origin of Species*[5] for almost a quarter of a century. Even then, he was hastily provoked into publication by the findings of another biologist, Alfred Russel Wallace, who had independently come up with exactly the same mechanism for natural selection and was potentially going to upstage Darwin.[6]

So, the popular idea that Darwin immediately recognised the significance of these rather drab little birds, and then used them as the basis for his world-changing theory, is palpably nonsense. In reality, the birds that enabled him to develop his ideas were far less exotic: the various breeds of domestic pigeon, along with another group of species he collected from the Galápagos, the mockingbirds.[7]

You may already be wondering why Darwin's finches appear in this book at all. Paradoxically, they do so precisely because of the myth of their apparently crucial importance to Darwin, which has taken such hold over time that evolutionary biologists have gone on to study these birds more closely than virtually any other group of living organisms. By doing so, Darwin's successors have made breakthrough insights into the nature and process of evolutionary change, as we shall see.

I shall go on to unravel the messy, complex yet compelling story of what actually happened during Charles Darwin's five weeks on the islands and, more importantly, what happened after he returned home. But first, what of the finches themselves?

Unlike the other chapters in this book, which tell the story of a single bird, this one focuses on at least fourteen, and possibly as many as eighteen, different species. That scientists disagree even as to exactly how many kinds of Darwin's finches there are tells us a lot about the fluidity and mutability of these birds – and also the fluidity and mutability of the very concept of a 'species'.

To keep things simple, let's start with the fourteen species that more or less everyone agrees on, thirteen of which are endemic to the Galápagos Islands, along with another, the Cocos Finch, found on the island of that name, some 750km to the north-east.

Despite the name, Darwin's finches are not actually members of the finch family (*Fringillidae*) at all. Neither are they members of the New World sparrows and buntings (*Emberizidae*), as they were regarded for more than a century after they were first discovered. Nowadays, it is widely agreed that they belong to the tanagers and their allies (*Thraupidae*), the largest New World family of songbirds, which, after much taxonomic to-ing and fro-ing, currently contains 383 extant species.[8] One of these, the dull-coloured grassquit, is likely to have been the ancestor of Darwin's finches. It lives in the subtropical lowland forests of western South America, and its drab appearance lives up to its name.[9]

This ancestral grassquit is thought to have arrived on the Galápagos some two to three million years ago, most likely carried there by a powerful storm. It eventually mutated and evolved into the maximum of seventeen species recognised as living on the islands today.[10] This happened via the process known as adaptive radiation, whereby, over time, a single ancestral species diversifies to fit different ecological niches, leading to a number of new and often very different looking, but nevertheless closely related, species.

If all the proposed 'splits' between the various species are accepted, those seventeen are the two warbler-finches (grey and green); the three tree-finches (large, medium and small), plus the woodpecker finch and mangrove finch; the vegetarian finch; the five ground finches (large, medium, small, sharp-beaked and Genovesa); three cactus finches (common, Genovesa and Española); and finally, the intriguingly named vampire finch, which, when food is scarce, survives by drinking the blood of the much larger Nazca boobies, using its sharp beak to pierce their skin.[*]

---

[*] Darwin's finches are currently placed in four genera: *Certhidea* (warbler-finches), *Camarhynchus* (tree-finches, woodpecker finch and mangrove finch), *Platyspiza* (vegetarian finch), and *Geospiza* (ground finches, cactus finch and vampire finch).

This may be the most dramatic example of a species evolving to exploit an available food resource, but each of Darwin's finches has also evolved to do so. As their names suggest, the two warbler-finches have a very thin bill, which allows them to probe into foliage to extract tiny insects. In contrast, the quintet of ground finches have thick bills, which they use to feed on seeds and beetles. And, rivalling the vampire finch for ingenuity, the woodpecker finch uses the spines of cactus plants as a tool, to probe into the holes of trees, and extract the invertebrates lurking within.

These differences in food and feeding methods – and especially the variation in the size and shape of the birds' bills – are crucial, because this is how new species evolve. Imagine a world where every species of bird lived in the same habitat and fed on the same things, caught or obtained in the same way. If that were the case, there would be very little variation in shape and structure between the different species.

Of course, that is very far from reality. We, and the birds, live on a planet where there is a huge variety of different foods and habitats, and therefore an almost infinite range of opportunities. That's why there are at least 10,700 (and probably far more) different species of bird, each of which has evolved to exploit a very specific ecological niche.[*]

Scientists call the observable differences between individuals within a population of organisms 'phenotypes' – defined by Oxford Dictionaries as 'the set of observable characteristics of an individual

---

[*] The single-volume *All the Birds of the World* (Lynx Edicions, 2020) includes every species accepted by any one of the four major 'world lists' of birds: a total of more than 11,500 different species.

resulting from the interaction of its genotype with the environment'. Thus, just as individual people have different eye colour or blood type, so birds may also vary from individual to individual. Many – if not most – of these differences do not matter. But when the environment changes – especially suddenly, perhaps as a result of a dramatic shift in climate – these tiny variations might end up making the difference between survival and oblivion.

That's exactly what had been happening on the Galápagos. Over time, Darwin's finches had adapted to the islands' 'unstable and challenging environment', resulting in noticeable differences in their body size, shape of their beaks, plumage, feeding behaviour and even their songs.[11]

So even though the various species may look different, behave differently and exploit very different ecological niches, they are all closely related, having originally evolved from that single common ancestor.

There is no doubt that Darwin did notice his eponymous 'finches' as he visited the various islands of the Galápagos archipelago – given that they are among the most common and ubiquitous of the very few songbirds present there, he could hardly have failed to.

In *The Voyage of the Beagle*, Darwin's account of the expedition published soon after his return, the contents page includes a passing reference to 'curious Finches' – along with other equally intriguing entries, such as 'Great Tortoises', and 'Marine Lizard, feeds on Sea-weed'.[12] But in more than 500 closely written pages, there are only another ten references to finches, all brief and lacking any detail, apart from this, with hindsight rather prescient, comment: 'Unfortunately most of the specimens of the finch tribe were mingled together; but I have strong reasons to suspect that some of the species of the sub-group *Geospiza* [the ground, cactus and vampire finches] are confined to separate islands.'[13]

In fact, although Darwin realised that the birds he and his servant, Syms Covington, were collecting were separate species, he initially believed that they were not related to one another at all. Instead, he thought that they came from a range of families, including the New World blackbirds (*Icteridae*), 'gross-beaks', warblers, wrens and true finches. He did notice that the mocking-birds he encountered on each island appeared to be distinct from one another. But at no point did he consider that the 'finches' and their allies shared a common ancestor. As he freely admitted, he did not even label them properly, so that when he returned to the UK a year later, in October 1836, he could not work out on which island he had collected any particular specimen. This did not quite render his collection completely useless, but it did not help.

But he then had a stroke of luck. In early January 1837, Darwin sent his specimens to the Zoological Society of London (ZSL), which in turn passed them on to its museum curator, the orni-thologist, taxonomist and bird artist John Gould, for closer examination. Less than a week later, at a meeting of the Society, Gould dropped his bombshell: the birds Darwin had considered to be a miscellaneous group of unrelated species were, in fact, from the same family. Gould described them as 'a series of ground Finches which are so peculiar . . . [that they represented] an entirely new group, containing twelve species'.[14] Later, he also discovered that the 'varieties' of mockingbird Darwin had collected were also distinct species, rather than mere races.*

This had huge implications for the scientific theories Darwin and others were beginning to develop. If these very different looking birds, each perfectly adapted to a specific ecological niche (notably through their varied feeding techniques), belonged to the same family, then they must surely have descended from a common ancestor. If that were the case, as it certainly appeared, then either the Creator was playing a clever trick on humanity, or one species really could – over time, and with the right environmental conditions – evolve into a suite of completely new species.

As an archipelago of remote, oceanic islands, far from any other landmass, and with a paucity of species, the Galápagos Islands were the perfect place for this transformation to occur. Because there were so few songbird species present, and therefore a number of vacant ecological niches available, the newcomers were able to exploit each specific niche without any competition. And yet, even when faced with this apparently overwhelming evidence for evolution by natural selection, Darwin appears to have dismissed it. True, in the second edition of *The Voyage of the Beagle* (now called *Journal of Researches*), published in 1845, he did present a detailed – and beautifully illustrated – description of the differences between what Gould had now decided were thirteen different species:

* Ironically, although John Gould's discovery that all these birds were closely related to one another was in some ways a catalyst for Darwin's later work, Gould himself was never a supporter of Darwin and went on to disavow his theories. See Jacqueline Banerjee, 'John Gould and Darwinism': https://victorianweb.org/science/gould/darwinism.html.

The remaining land-birds form a most singular group of finches, related to each other in the structure of their beaks, short tails, form of body and plumage . . . The most curious fact is the perfect gradation in the size of the beaks in the different species of *Geospiza*, from one as large as that of a hawfinch to that of a chaffinch, and (if Mr Gould is right in including his sub-group, *Certhidea*, in the main group) even to that of a warbler.[15]

Summing up his findings, Darwin comes to what appears to be a startling conclusion: 'Seeing this gradation and diversity of structure in one small, intimately related group of birds, *one might really fancy that from an original paucity of birds in this archipelago, one species had been taken and modified for different ends.*' [My italics]

Yet in 1859, when he finally published *On the Origin of Species*, he did not mention the finches even once. Towards the end of his life, in 1877, he did admit that, at the time, his own religious beliefs would not have entertained the heretical concept of evolution by natural selection:

When I was on board the *Beagle*, I believed in the permanence of species, but, as far as I can remember, vague doubts occasionally flitted across my mind. On my return home in the autumn of 1836 I immediately began to prepare my journal for publication, and then saw how many facts indicated the common descent of species, so that in July, 1837, I opened a note-book to record any facts which might bear on the question. But I did not become convinced that species were mutable until I think two or three years had elapsed.[16]

In rejecting the very idea of mutability, Darwin was simply adhering to the principles espoused by the Church of England, into which he had been born and raised. Going against these would be unthinkable (see chapter 4). Later on, however, Thomas Henry Huxley (who became known as 'Darwin's Bulldog' for his staunch defence of Charles's theories) denounced the Church as 'the one great spiritual organisation which is able to resist, and must as a matter of life and death, resist, the progress of science and modern civilisation'.[17]

Neither, by the way, did Darwin, Huxley, or any of their contemporaries ever actually call these birds 'Darwin's finches'. That now famous label was first coined in 1935, more than half a century after Darwin's death, by Percy Lowe, the curator of birds at London's Natural History Museum. While giving a talk to his fellow scientists, Lowe explained his choice of the new epithet: 'We know that it was the diversity presented by these Finches, as well as the Mocking-birds, tortoises, and plants, which started Darwin down that brilliant corridor of thought which led to his conception of the origin of species.'[18] If any one individual can take the credit for launching the enticing and all-pervasive – yet utterly false – myth about Darwin's 'eureka moment' on encountering these little birds, it is Mr Lowe.

Two historians of science, Frank Sulloway in the US in the early 1980s, and more recently the Singapore-based Briton John van Wyhe, have delved into both the origins and the sheer persistence of the mythology surrounding Darwin's finches, and in particular the undue – and indeed misleading – importance it has been given over time. Van Wyhe points out that Darwin never specifically cited either the Galápagos or the finches as the inspiration for his later theories. Neither did a single one of the 'torrent of obituaries' after Darwin's death in 1882 mention the finches, with few even acknowledging the supposed importance of his visit to the islands.[19]

The first reference to the idea that Darwin's experiences on the islands were linked to his theory of evolution by natural selection did not appear until the centenary of his birth, in 1909. That was when his son Francis – himself a distinguished botanist – made the first explicit connection between the two:

> 'Darwin's attention was "thoroughly aroused" by comparing the birds shot by himself and by others on board. The case must have struck him at once – without waiting for accurate determinations – as a microcosm of evolution.'[20]

This appears to have been the tipping point: the origin of the myth, after which the false connection between the finding of the finches in the Galápagos, and Darwin's development of his later theories, became more and more pervasive. A quarter of a century later, when Percy Lowe coined the term 'Darwin's finches', the connection was already well and truly entrenched in popular culture.

In a further delicious irony, it was that false connection – and the supposed importance of the finches to Darwin – that ultimately sparked off events that would, over the course of the second half of the twentieth century, lead scientists to an extraordinary conclusion. This was that, even if Darwin's finches were not the example that inspired the man himself, they are still one of the best demonstrations not just of Darwin's own theory, but of those developed by his successors.

The first man to study Darwin's finches more closely in the field began his career as a humble schoolmaster, yet ended up as one of the greatest evolutionary biologists the world has ever known: David Lack.

Today, David Lack is revered in academic circles as the 'father of evolutionary ecology', as the subtitle of a recent biography describes him.[21] But in the years leading up to the Second World War, he was a Devon schoolmaster and largely amateur ornithologist.[22]

Even though Percy Lowe had first drawn attention to Darwin's finches, he remained unconvinced that they genuinely were a prime example of Darwin's theory. To find an answer to this conundrum, the ZSL's secretary, Julian (later Sir Julian) Huxley, proposed that Lack should be despatched to the Galápagos to study the birds in the field, and so prove – or disprove – their relevance once and for all. Thus, in early November 1938, David Lack and a small party of fellow scientists left Liverpool on the long sea journey to the other side of the world.[23] Almost six weeks later, the party finally disembarked on Chatham Island in the Galápagos – coincidentally, the very same place where Darwin himself had first made landfall just over a century earlier.

Lack spent almost five months living on the islands, studying the birds more closely than anyone had done before. The following April he set sail with his precious cargo of live finches, which he and his colleagues had trapped on different islands (unlike Darwin, keeping a careful record of their places of origin). He did not, however, return directly home, but instead went to San Francisco, as he feared the birds would not survive the much longer sea voyage to England.

Lack remained in California for four months, making careful notes and measurements of the finches, before heading to New York on his way home. It was there, on 3 September 1939, that he heard the news that Britain was at war with Germany. Fortunately, later that month, he did manage to get a safe passage on a ship back to Britain.

A year later, in 1940, David Lack submitted the first of two major works on the finches: a paper entitled 'The Galápagos Finches (*Geospizinae*), A Study in Variation'.[24] This came up with some surprising conclusions: according to Lack, the obvious variations between the size and shape of the different species' bills were not connected with their food, feeding habits or lifestyle. Instead, he suggested, they were simply a way for each species to be reproductively isolated – and so avoid interbreeding with – the others.

Because of the delays caused by the war, Lack's paper was not published until 1945. By then, as his biographer Ted Anderson reveals, his theories on the evolutionary history of the various species of finch had undergone a more or less complete U-turn.[25]

In 1947, after a rejection from his usual publisher H. F. & G. Witherby, Lack published the book that would seal his reputation as a scientist: *Darwin's Finches: An Essay on the General Biological Theory of Evolution*.[26] In this seminal work, he repudiated his previous theory, and instead proposed that almost all of the variations between the different species were indeed the result of the pressures of natural selection. These, he argued, had given rise to adaptive radiation, and the evolution of entirely new species.

These changes were, he believed, driven first by the birds being geographically isolated from the South American mainland, and then by becoming ecologically isolated, i.e. exploiting different food resources and habitats on the various islands of the Galápagos. This process would allow the three different species of ground finch – small, medium and large – to co-exist on the same island.

David Lack's breakthrough theories took more than a decade to be fully accepted by the scientific community. He eventually went on to have a distinguished academic career, as head of Oxford University's world-leading Edward Grey Institute of Field Ornithology. Lack continued his fieldwork on birds in the UK and abroad, inspiring many of today's top scientists, until his untimely death, aged just sixty-two, in 1973.

In John van Wyhe's exemplary study of the history of the false cultural connection between the Galápagos finches and Darwin's famous theory, he concludes by outlining a six-stage process, which he calls (with a nod to Darwin himself) 'the evolution of these legends' through time:

1. The Galápagos were never mentioned.
2. They were mentioned but no special role given to them.
3. The theory was later derived from them.
4. The theory was conceived on them.
5. The finches played a special role.
6. The finches caused the theory to be conceived on the Galápagos.

As he points out, this cumulative error is not because of some grand design or sinister conspiracy; it is simply 'the result of countless individuals' idiosyncratic takes on what they had heard and read, in addition to their political or personal motives on spinning the story of Darwin in a certain way at particular times.'[27] Even the BBC Natural History Unit – and the great Sir David Attenborough – fell for this false narrative, when, in his celebrated 1979 TV series *Life on Earth*, he stated that 'Darwin had noted similar variations in the bills of the finches of the Galápagos Islands and regarded them as powerful evidence for his theory of natural selection.' Another BBC TV series, the BAFTA-winning drama *The Voyage of Charles Darwin*, broadcast the previous year, also mixed myth and reality, in this fictionalised first-person account:

> But of all the creatures that excited my curiosity on that extraordinary group of islands, the ones that were to influence the development of my theories over many years were so dull and common in appearance, that I very nearly failed to notice them at all. These were the finches, that climbed and darted about amongst the cactus-trees and other bushes.
>
> Although they were all clearly related to each other, there were also distinct differences, particularly in the size and shape of their beaks.
>
> The true significance of this, and the possibility that it might shed some light on their distribution throughout the islands, did not occur to me; and those that I kept for my collection became mingled together in the most haphazard fashion.[28]

Despite this accidental, almost random development of the legend, though, it does turn out to have had a major influence on our understanding of the mechanisms of evolution.

A husband-and-wife team of biologists, Peter and Rosemary Grant, have closely studied the finches of the Galápagos for more than four decades. Their painstaking, groundbreaking work has shown that Darwin's finches can indeed illuminate exactly how new species evolve. The story of how they did so is one of the most unexpected of all tales of scientific discovery.

As biological science becomes ever more complex, many breakthroughs – especially those regarding the evolutionary relationship between different species of birds – are now made in laboratories, rather than in the field. The biologists' latest tool is the chemical analysis of DNA which has been extracted from the blood or feathers of a dead or living bird. The scientists see how the strands of DNA from different species match up against one another – the nearer the match, the more closely related the species are to one another. This revolutionary technique, then known as DNA–DNA hybridisation, was originally developed during the last decades of the twentieth century, by US scientists Charles G. Sibley and Burt L. Monroe. Though considerably modified and improved since, it has led to radical changes in our understanding of the relationship between different species, with groups previously thought to be close relatives discovered to be completely unconnected, and vice versa.[29]

With this shift towards lab-based discoveries, it might be thought that the era of the field biologist – fit and tanned, with their sleeves rolled up and ready for action – would be over. Yet the life's work of Peter and Rosemary Grant refutes that idea.

Peter Grant was born in London in October 1936; his wife Rosemary was born in Westmorland (now Cumbria) that same month. From early childhood, Peter was fascinated by natural history, although his cousin, the veteran travel and nature writer Brian Jackman, recalls an important early difference between them: 'Peter . . . shared my love of the natural world, although even then we viewed it in a different light. Already he was beginning to take a more scientific approach, while I remained an incurable romantic.'[30]

After graduating – Peter from Cambridge, Rosemary from Edinburgh – they each embarked on an academic career in evolutionary biology. Having met in 1960 at the University of Vancouver, they married two years later, and decided to make North America their home. In 1985, after a period at McGill University in Montreal, Canada, the Grants moved to Princeton University, New Jersey, where they have worked ever since.

Actually, that is not quite true. The Grants may have been based in North America, but since 1973, when they first visited the Galápagos to study Darwin's finches, they have spent more time there than they have at home. They have done so while raising two daughters – Nicola and Thalia – who not only lived on the islands for much of their childhood, but also contributed to their parents' pioneering work.[31]

When the Grants first arrived on the Galápagos, they wanted to investigate the relationship between the various species of Darwin's finches; a task they thought might take a year or more. Almost half a century later, well into their ninth decade, they are still obsessed with these curious little birds: 'Darwin's Finches are unique in what they offer biologists. They are so similar to each other that transformation of one species into another can be reconstructed easily. They are accessible; their behaviour can be studied easily because they are tame.'[32]

Although they no longer spend long periods doing fieldwork, Peter and Rosemary still make periodic trips to the Galápagos, specifically the tiny island of Daphne Major, the site of their long-term field study of the finches. (Ironically, Charles Darwin never visited – even saw – this particular island for himself.) With a total surface area of less than half a square kilometre – roughly the same as five standard-size football pitches – this inaccessible, rocky island is in many ways the perfect site for studying a self-contained, sedentary and very tame cohort of birds, by getting to know them as individuals. Yet in other ways, the island is far from the ideal study site. Living conditions on Daphne Major are brutal. The Grants were cut off from civilisation for months on end, and had to be entirely self-sufficient in both food and water. Another major issue is the unpredictable and extreme climate, ranging from long, dry spells to weeks of drenching rain. This is especially true during the climatic phenomenon known as El Niño (the Christ Child), when a warming of the Pacific Ocean brings unseasonal, persistent and often very heavy rainfall to the islands, interspersed with unusually long periods of drought.

The weather caused major discomfort for the Grant family, making studying the birds even more difficult. Yet its unusual extremes would prove to be the key to their discoveries about evolution in Darwin's finches; evolution which at times seemed to be taking place not over centuries or millennia, but in real time, in front of the astonished observers, as one profile of the Grants recorded. 'Peter and Rosemary Grant are members of a very small scientific tribe: people who have seen evolution happen right before their eyes. For the Grants, evolution isn't a theoretical abstraction. It's gritty and real and immediate and stunningly fast.'[33] The key to this rapid and unprecedented change was, it turned out, the extremes of climate that at the time were making living and working on Daphne Major so challenging.

Peter and Rosemary Grant's work was first brought to public attention in a Pulitzer Prize-winning book, *The Beak of the Finch*, by the US writer Jonathan Weiner, published in 1994.[34] Subtitled 'Evolution in Real Time', it tells a gripping tale. Weiner describes how, after long periods of drought, brought about by unusual climatic events in the Pacific, the species the Grants were studying – a sub-group of Darwin's finches – began to evolve into new and different species before their very eyes.

This rapid 'evolution in action' was quite contrary to what everyone – including the Grants – had assumed prior to their first visit to the Galápagos. The generally accepted consensus was that Darwinian evolution proceeded at a snail's pace, over vast aeons of time, as tiny morphological and behavioural differences between individual creatures slowly and gradually pushed them apart, eventually allowing new species to evolve. As Weiner explains, this meant that when the Grants first arrived on Daphne Major, in 1973, they were simply hoping to get a 'snapshot' of the current evolutionary status of the various different species of finch, notably the medium and small ground finches: 'Watching these birds would be like watching the stars for an astronomer, or the mountains for a geologist. Even one hundred years in the Galápagos would be a snapshot.'[35]

Even during that very first visit, however, something about these little birds suggested to Peter and Rosemary that they should extend their study period on the island. Coincidentally, a radical new theory had emerged just a year earlier: that of 'punctuated equilibrium', proposed by two young US scientists, Niles Eldredge and Stephen Jay Gould. This challenged the widely held belief that evolution occurred smoothly, gradually and very slowly, replacing it with a more dynamic model, in which long periods of stasis – where little or nothing changes – are interrupted by short periods when evolution can, and does, occur very rapidly indeed.[36] Yet, as Peter Grant notes, although they had heard of Gould and Eldredge's theory, they were not especially influenced by it.[37]

Peter does point out that scientists had long been aware of very rapid evolution in simple organisms such as microbes in hospitals – leading to antibiotic resistance – and agricultural pests. They were also familiar with the celebrated story of the peppered moth: how, during times of industrial pollution, when soot coated the trees where the moths perched, darker forms were less likely to fall victim to predators, and so these rapidly spread through the population. But, as he says, 'the idea of long-lived animals such as birds and mammals evolving observably in our lifetime was not prevalent when we began, nor was the prospect of studying and measuring it.'[38]

For the first four years or so of their study, the rains on Daphne Major came and went as predicted, with the dry season followed, as usual, by the wet season. With an abundance of easily obtainable food, mainly in the form of seeds, the finches flourished, their numbers rising each wet season and then falling back again during the dry season. The Grants' study was going to plan.

Then, halfway through 1977, things changed. The predicted rains, which until then had been as regular as clockwork, were a mere fraction of their usual amount – about 24mm in total. As the drought worsened, so the birds began to struggle. They did not even bother to mate, let alone build nests, lay eggs and raise a family.[39]

Yet some birds did somehow manage to find food in these harsh conditions. When the smaller seeds became scarce, some of the medium ground finches, which had slightly larger beaks than their peers, were able to feed on the larger seeds, which were still available. As a result, these birds survived better than their smaller-billed counterparts, which starved to death. The larger-billed birds then passed on this lifesaving characteristic to their offspring, through genetic inheritance.

But smaller birds of the same species, and those of another species – the small ground finch – whose beaks were not as large or strong, simply could not crack open the larger seeds, and so were forced to feed on the smaller seeds of a different plant. Unfortunately for them, that particular plant produces a sticky substance, which caused their head feathers to become matted, exposing the skin beneath to the baking hot sun. In many cases, this resulted in the smaller birds dying. Others simply perished from starvation.

By the time the Grants returned home, in December 1977, the

total population of Darwin's finches on Daphne Major had fallen by almost 90 per cent, from 1,400 individuals in March 1976 to just 200 survivors less than two years later. The commonest species on the island, the medium ground finch, had seen its population crash to a fraction of its former numbers. The small ground finch – of which just a dozen had formerly been present – fared even worse; down to a single, lonely individual. For the Grants, this unforeseen and unexpected disaster for the birds presented a tantalising opportunity, motivating them to continue their studies, to discover what would happen next.

Two of Peter Grant's PhD students from McGill University, Peter Boag and his wife Laurene Ratcliffe, had remained on the island during and after the drought year, making valuable observations. Towards the end of the drought, the entire Grant family, including Nicola and Thalia, followed up this fieldwork by methodically recording which individuals had managed to survive, and which had died.[40]

Back home in Canada, the Grants and Peter Boag sifted methodically through their carefully obtained data. They focused on measuring the beaks of the birds that had died during the drought and comparing those measurements with the beaks of the birds that had managed to find enough food to survive. That was when they made an astonishing discovery: one that would revolutionise what scientists thought they knew about the process and timing of evolution by natural selection.

The differences were, to an outsider, so minuscule they seemed hardly worth mentioning. The specimens of the medium ground finch that had survived had beaks that, on average, were between five and six per cent larger than those that had died. Given the small size of the birds, this difference was measured in fractions of a millimetre – far too small to be seen by the human eye. Yet the Grants realised that for the individual birds, those tiny variations made the difference between life and death. This was a major scientific breakthrough. 'Not only had they seen natural selection in action', writes Weiner: 'It was the most intense example of natural selection ever documented in nature.'[41]

In the long term, provided the climate soon went back to normal, the survival of the larger-billed – and larger-bodied – birds might not make much difference, but for one other crucial factor. Because male medium ground finches tend to be slightly larger than their mates, far more males than females had managed to survive the drought. This meant that when the rains finally returned in January 1978, six months later than usual, a new factor – sexual selection – came into play.

With just one female for every six males, the females found themselves in a 'buyer's market': they were able to pick the 'best' males from the surviving cohort. And that's exactly what they did: choosing to mate with the largest males, with the biggest and strongest bills. Thus, the potentially temporary shift towards larger-billed birds soon became permanently entrenched in the population.

In less than a year, a major evolutionary change had occurred – and more importantly, it had been seen to occur. Peter and Rosemary Grant had found their life's work, and no-one would ever be able to look at evolution in quite the same way again.[42]

*

Darwin's finches may be the clearest – and are certainly the most famous – example of adaptive radiation. But they are far from the only one. Several other similar examples also occur on isolated oceanic islands. That's because these remote locations usually have a paucity of native songbird species, yet often still have a wide range of vacant habitats and ecological niches, so are the ideal location for this process to happen.

In the forests of Madagascar, the twenty-one species of vangas[*] are a diverse group of medium-sized, shrike-like songbirds. Vangas are thought to have evolved from a single common ancestor, which reached the islands – almost certainly from the mainland of Africa – more than 20 million years ago.[43]

Like Darwin's finches, vangas have diversified to exploit different food sources. Consequently, they have evolved over time into a wide range of different-looking species, which – again like the finches – have very different-shaped bills. This inevitably leads to very different feeding habits. The nuthatch vanga, as its name suggests, behaves rather like a nuthatch (though it cannot climb down tree trunks); the helmet vanga uses its huge, grosbeak-like bill to crush beetles and lizards; most remarkably of all, the sickle-billed vanga deploys its long, thin, curved bill to prise insects out of their hiding places beneath the bark of a tree. Again, just like Darwin's finches, to the untrained eye these closely related species appear to belong to completely different families.

On another major oceanic island group, Hawaii, the Hawaiian honeycreepers have undergone a similar evolutionary process,

[*] Members of the family *Vangidae*.

leading to more than fifty different species, which are able to feed on a wide range of foods including fruit, seeds, nectar and invertebrates. Like both Darwin's finches and vangas, Hawaiian honeycreepers evolved from one or more finch-like species, which arrived on these isolated islands between four and five million years ago.[44]

The sheer diversity of bill size and shape among the Hawaiian honeycreepers is, quite simply, astounding. By evolving over time to exploit the many different ecological niches on the islands, they have replicated almost all of the very varied bill shapes of the 6,000 or more species of songbird in the world.[45]

Ironically, however, it is that very diversity, and ability to specialise, that has led to these island species becoming threatened by habitat loss and climate change. Another major problem is that, by definition, island birds tend to have small populations and restricted home ranges, and so when conditions change they are disproportionately threatened with extinction, especially when compared with similar species that inhabit larger landmasses (see chapter 4). Nowhere is this more obvious than the fate of the Hawaiian honeycreepers, of which more than half the known species have now gone extinct – many during the past few decades.[46] These include the Kaua'i akialoa, which went extinct in 1969, the Maui 'akepa, which disappeared in 1988, and the po'o-uli. Having only been discovered in 1973, the po'o-uli was declared extinct less than fifty years later, in 2019, following the last confirmed sighting in 2004. Before then, this snail-eating honeycreeper had earned the unwanted distinction of being 'the world's rarest bird'.[47] Today, only its euphonic Hawaiian name survives, the po'o-uli having become, like the dodo, a mere ghost of its former self.

Just as the fifty or so species of Hawaiian honeycreepers are an example of adaptive radiation among birds, so passerines as a whole also demonstrate this process, but on a much larger scale. And now two books – one a scholarly tome, the other a work of popular science – have made us think again about exactly how bird species evolve.

The more recent of the two, *The Largest Avian Radiation*, maps the process by which well over half the world's bird species – the perching birds of the Order *Passeriformes* – evolved.[48] It presents a new and revolutionary classification, said by the editors to be a 'unifying theory to explain how the prodigious variation of Earth's biodiversity is generated'. This is based on more than a thousand scientific papers, which the editors have ably synthesised into a single, albeit rather weighty, volume.[49]

The scientists whose work is presented here have used new and more precise molecular methods, which produce far more accurate comparisons between species and families than ever before. In doing so, they have dropped a metaphorical bomb into the accepted system of classifying birds, overturning many long held and widely accepted beliefs. Some families have been split apart, while others have merged, creating a plethora of new connections. Of these, one of the most fascinating – and to traditional birders, surprising – examples is that of the Old World family known as the 'warblers'.[50]

The current official 'British List' of birds recorded in a wild state in Britain,[51] produced and updated by the British Ornithologists'

Union (BOU), includes a total of fifty-four species of warbler. These range from the common and familiar chiffchaff, black-cap and whitethroat to vagrants such as the olive-tree, eastern crowned and pale-legged leaf warblers, which have only occurred a handful of times in Britain. Take a closer look, though, and you discover something rather odd. Cetti's warbler, a skulking species found in marshlands across southern Britain, is separated from the other fifty-three warblers by the familiar – and supposedly unrelated – long-tailed tit. You might be forgiven for wondering whether this means that Cetti's warbler is not a warbler at all, or the long-tailed tit *is* a warbler, or that neither (or perhaps both?) are warblers.

When you consult the checklist of the birds of the 'Western Palearctic'[52] – the biogeographical area that includes Europe, North Africa and the Middle East – things become even more puzzling. Here, unlike the BOU Checklist, the species are separated into families. And what used to be a single family of warblers has now been split into no fewer than six.

Interspersed among these new families are several others, not usually regarded as related to warblers at all: the bulbuls (*Pycnonotidae*); the aforementioned long-tailed tit (placed in its own, single-species family, *Aegithalidae*); the bearded reedling (also in its own family, *Panuridae*); babblers (*Leiothrichidae*); and most surprisingly of all, the larks (*Alaudidae*) and the swallows and martins (*Hirundinidae*).

This radical new classification mirrors chapter 11 of *The Largest Avian Radiation*. Titled 'Superfamily Sylvioidea: The Old World Warblers and Their Allies', it contains no fewer than 1,200 different species – roughly one in five of all passerines, and more than one in ten of all the world's birds.[53]

If these findings are correct – and there is no reason to suppose they are not – then the traditional view that 'warblers' all belong to a single family, a view that has prevailed since the days of Gilbert White in the late eighteenth century, is no longer tenable.

If that seems rather mind-boggling, we should remember that any attempt to classify the world's birds – or other living organisms – has always been subject to uncertainty. Even with the most cutting-edge science at our disposal, until recently we could not say exactly how today's assemblage of bird families and species actually evolved. Now, however, the new laboratory techniques are giving us definitive answers to the relationship between families and species.

Classification has always changed over time. Many of those bird books from the early days of ornithology show a very different system from the one we use today. For example, they might classify all 'birds of prey' – including owls, as well as hawks, kites, eagles, harriers and falcons – as 'rapacious birds'.

Superficially, this makes sense: both hawks and owls have sharp talons for killing their prey, and a hooked beak for tearing it apart to feed. Yet today we know that these similarities are purely circumstantial: the result of a process known as convergent evolution. This occurs when two totally unrelated species share a similar ecological niche and, as a result of their habits, develop similar physical characteristics.

Today we may be amused by the notion, held by those pioneering ornithologists, that hawks and owls were thought to be related to one another, just because they have similar beaks and claws. Yet, until very recently, although hawks and falcons were placed in two different orders (*Accipitriformes* and *Falconiformes*), they were generally considered to be related. Now, thanks to developing scientific knowledge, we know that falcons are no more related to eagles and hawks than owls are; in fact, they are nearer to parrots and passerines.

Nevertheless, despite the recent breakthroughs outlined in *The Largest Avian Radiation*, birders will no doubt continue to use the word 'warblers' to describe the small, insectivorous and mostly migratory birds they see on their travels. Over time, later generations may come to see them in a different light and wonder why we ever thought they were closely related to one another.

The other book that challenged the way we view the world's songbirds is a much slimmer volume than *The Largest Avian Radiation*, yet it tells us more about our cultural attitudes to birds – and to different nations – than any other publication in the last decade.

Where Song Began: Australia's Birds and How They Changed the World[54] was published in 2014 by the Australian biologist and science writer Tim Low. It was, perhaps surprisingly for such a complex subject, both a critical and commercial success, reaching the bestseller lists and becoming the first nature book to win the prestigious General Non-Fiction prize at the Australian Book Industry Awards.

Low's compelling book explains how scientists have overturned long held assumptions about how Australasia's diverse cohort of songbirds evolved, making us look at the evolution and spread of birdsong throughout the world in an entirely new way. As the leading Australian birder Sean Dooley wrote in his *Sydney Morning Herald* review, Low's 'brilliantly readable book . . . not only gives Australian birds recognition long overdue, but allows for a fresh understanding of the way the world (and particularly our island continent) functions'.[55]

What Tim Low did was to confront another 'sacred cow' of ornithology. Until very recently, it was believed that birds evolved in the northern hemisphere, gradually spreading out from the vast Eurasian landmass until some species finally managed to reach the remote land of Australia where, having been isolated, they continued to evolve into new and different families and species. This theory was promulgated by none other than the great German-American evolutionary biologist Ernst Mayr, known as 'the Darwin of the twentieth century'. He argued that different bird families first arrived in Australia in separate waves, and then gradually evolved into the range of species found there today. But what Mayr failed to realise was that many Australian birds that looked and behaved the same way as Eurasian species were not closely related to them at all, but were instead another series of examples of convergent evolution.

In fact, as Tim Low reveals, more or less the opposite scenario from that proposed by Mayr was the case. More than half the world's birds – including parrots, pigeons and passerines (song-birds) – had actually evolved in that vast prehistoric supercontinent known as Gondwana. It was from here, some 180 million years ago, that several landmasses we know today, including South America, Africa, Arabia, the Indian subcontinent and Antarctica, as well as Australasia, split off and drifted apart. Some time afterwards, birds that had evolved in Australia began to spread into various landmasses, including Asia, eventually colonising the entire world. Some, such as the finches and ravens, eventually returned, millions of years later.

Why this theory took so long to be accepted is the central question of Tim Low's book. In his introduction, he explains that what he calls 'northern orthodoxy' – a view that Australia and its wildlife are somehow 'second best' – is hard-wired into the way Europeans and Americans regard his nation, and also, ironically, how Australians have sometimes viewed themselves. As he points out, this prejudice goes back all the way to the European settlers in the late eighteenth century, and their antipathy towards the unusual fauna they found there:

> The egg-laying platypus and echidna and quirky marsupials were taken to be primitive mammals that had survived only because Australia was isolated from other continents. This thinking did not stop Australians from feeling pride in their wildlife, but it was tempered by a belief that the mammals, and to some extent all the animals, were backward. The birds fitted this picture by seeming to lack any aptitude for song.[56]

Yet that inferiority complex could not cloud scientific enquiry for ever. By the 1970s, a period that also saw a global rise in the appreciation of Australian writing, art, film and music, scientists were at last beginning to question the orthodoxy so firmly championed by Ernst Mayr and the establishment. The

US scientist Charles Sibley used his new and iconoclastic DNA–DNA hybridisation technique to reveal that many of Australia's most familiar species, including robins, flycatchers, warblers and babblers, were not members of the Old World groups that their names indicated, but were more closely related to one another than to any bird families outside Australia.

Just as the American robin is not a robin, but a thrush, so these birds were hiding in plain sight, behind a false and misleading name: the result of homesick British settlers calling the new birds they saw after familiar ones from back home. The same applies to the Australian magpie and the Willie wagtail – named after the Eurasian magpie and pied wagtail, because of their superficial similarity in appearance or behaviour.[*]

Sibley went even further. These birds had not just evolved in Australia, he suggested, but their descendants had then colonised the rest of the world. This was incendiary stuff, as Low explains. 'Not only had Australia evolved its own songbird clade [a group of organisms believed to comprise all the evolutionary descendants of a common ancestor], it had been an exporter on a grand scale. Many of Europe and America's feathered finest, including magpies, jays and shrikes, were part of the Australian radiation. They are closer to Australian birds than to most northern birds.'[57]

Two other Australian biologists, Les Christidis and Dick Schodde, then took Sibley's work on another stage, creating a 'family tree' that placed two iconic Australian bird families

---

[*] Incidentally, not all Australasian songbirds are only found on that continent: orioles, crows and ravens, swallows and finches are from the same families as their Old World relatives. But the vast majority are, indeed, unique.

– lyrebirds and scrub-birds – as 'sister-families' to all other song-birds. The implications of this revelation were both astounding and, to the British and American mind, shocking. What Christidis and Schodde's work implied was that many celebrated garden or backyard birds – from the robin, blackbird and song thrush in Britain, to the cardinal and American robin in North America (as well as the Asian and African babblers, sunbirds and weavers) – originally evolved from Australian ancestors.

Inevitably, at first many western scientists flatly refused to accept such an iconoclastic conclusion. Even when, in 1991, the results were published in *Ibis*, the august journal of the BOU, Christidis and Schodde were forced to hedge their conclusions with the word 'hypothesis', suggesting that the theory might not actually be true. Only in 2004 was it finally proven that songbirds did indeed evolve in Australia, millions of years ago.[58] The old, outdated idea that Australia's fauna and flora were inferior to that of the rest of the world, a wrongheaded notion largely promulgated through western snobbery and colonialism, had at last been laid to rest.

With all these rapid developments in the science of evolutionary biology giving us new, exciting and sometimes discomfiting insights into the relationships between different bird species and families, it would be easy to forget about the fate of the group of birds that is more closely linked with evolutionary theory than any other. Yet Darwin's eponymous finches now face a whole range of threats to their continued survival and existence, not least, ironically, because of those unexpected climatic changes which revealed so much to Peter and Rosemary Grant about how evolution works.

As well as the climate emergency, and its unpredictable effects on the Galápagos Islands' weather and food supply, the finches are also having to deal with habitat loss, introduced predators, avian malaria and an invasive species of fly, *Philornis downsi*, which parasitises the birds by attacking their young while they are in the nest, in many instances killing them.[59] It is too soon to know whether conservation measures to eradicate predators and parasites will work, but what we do know is that if they do not, during the next half century some species of Darwin's finches are likely to go extinct.

Any loss is a disaster, but the extinction of these birds, which have taught us so much about the mechanics of evolution and changed the way we see all life on Earth, would be an even greater tragedy, to match that of the great auk, passenger pigeon and dodo.

6

GUANAY CORMORANT

Leucocarbo bougainvillii

Whenever I hear anyone arguing for slavery, I feel a strong impulse to see it tried on him personally.

Abraham Lincoln, Speech to One Hundred Fortieth
Indiana Regiment (1865)

As the sun sank towards the Pacific Ocean, the men could look forward to a few hours' respite from their back-breaking labours. Since well before dawn, they had been hard at work, following the same routine as they did for twenty hours a day, six days a week, throughout the year.

Early the next morning, long before sunrise, they arose from the thin grass mats they used as beds, and began work in darkness, using picks and shovels to remove their quarry from the hills. First, they loaded the heavy and malodorous substance into wheelbarrows, which they pushed to the tops of the cliffs around the edge of the island. Then, they emptied the barrows into canvas pipes, down to the barges beneath, to be ferried to the cargo ships offshore. After this, they went back to mining once again. Each man had to produce five tonnes – up to a hundred barrows-full – per day. If they refused, or disobeyed their orders, they would be summarily shot.

For sustenance, each worker was given a daily portion of rock-hard bread, dried meat and rice, usually infested with maggots. There was no fruit or vegetables, so many suffered from scurvy. They also endured conjunctivitis and a range of respiratory diseases, brought about by breathing the fine dust which rose constantly from the ground into their lungs.

At the end of yet another day, as they gazed towards the setting sun, did they realise that half a world away, that same sun was rising over their homeland? Months or years earlier, they had left China to escape famine, war and poverty, and start a new life in a new land: North America. They had been lured by the tantalising prospect of taking part in the California Gold Rush, where they would earn enough money to send back to their poverty-stricken families at home.

The reality was very different. At the port of Macao, hundreds of them had been crammed into the holds of rickety, rat-infested and dangerously overcrowded ships, before embarking on what seemed like an endless journey across the vast, barren emptiness of the Pacific Ocean. They spent almost the entire time in virtual darkness, struggling to breathe the acrid air, laden with the stench of their fellow travellers.

On some voyages, as many as one in four men did not survive, dying from disease, malnutrition, or violence from their fellow passengers or crew. Others jumped overboard, hanged or stabbed themselves, rather than face the horrors they knew would greet them when they finally reached the end of their journey.

Between 1850 and 1874, an estimated 87,000 Chinese workers – known as 'coolies', from an Indian word for hired labour – did manage to make it to their final destination. But this was not California, where they had been told they were heading. Instead, they found themselves far to the south: on the Chincha Islands, off the coast of Peru.

Once they arrived, they traded in one form of hell for another: the arduous work of obtaining one of the most valuable commodities in the entire history of global trade. Not the gold they had imagined they would be mining – but the droppings of seabirds.

When visiting South America, the world's most bird-rich continent, few birders bother to seek out the guanay cormorant.[*] After all, with so many exotic species, including dozens of colourful toucans, tanagers, manakins, parrots and hummingbirds, why would you go out of your way to look for such a common, and apparently ordinary, seabird? When I travelled to Peru, however, in May 2017, the guanay cormorant was one species I was really hoping to see. Not for its rarity or beauty, but because of all the world's 10,000-plus different birds, it has one of the most extraordinary stories to tell.

This is a tale of greed and profit, horror and hardship, vast riches and almost unimaginable suffering. It also changed the rural landscapes of Britain, Europe and North America, and transformed the way we grow and produce our food. It led to the guanay cormorant being dubbed 'the billion-dollar bird' – the most valuable wild bird in human history.

On my very first morning in Peru, I headed to Playa Pucusana harbour, an hour or so south from the capital, Lima. There, among the colourful fleet of fishing boats, I scanned through flocks of bulky Peruvian pelicans, elegant Inca terns and scavenging turkey vultures, searching for a medium-sized cormorant with a distinctively pied plumage. After a few minutes, I found what I was looking for. The bird, a smart adult, was perched on the rocks at the edge of the harbour, showing off its pale pink feet and deep crimson eyepatch. Its jet-black back, contrasting with the pure

---

[*] *Leucocarbo bougainvillii*, named in 1837 after Vice-Admiral Louis Antoine, Comte de Bougainville (1729–1831), a navigator who explored the South American coasts, and also gave his name to a well-known flower, bougainvillea.

white breast, made it look rather dapper, as if the bird had dressed for dinner.

Even though most visiting birders ignore the guanay cormorant, other people have long understood its worth. Its value is commemorated in that strange name, for the guanay cormorant is one of only a handful of the world's 10,000-plus bird species to be called after its most important by-product.* The name comes from the bird's droppings, known by a word derived from the indigenous Quechua language of Peru: guano.

The guanay cormorant is a member of one of the world's most cosmopolitan bird families, the cormorants and shags.† There are roughly forty species in all, which between them live and breed on all seven of the world's continents.

Guanay cormorants can be found along the Pacific coast of South America, breeding from near the equator in the north to Chile in the south. A relict population did survive for a while on the Atlantic coast of Argentina, but now appears to have died out. Outside the breeding season, the birds wander further afield, dispersing north to Central America and south to Cape Horn, especially during years when the disruptive El Niño weather system is dominant.

---

* The only other species whose official English names commemorate something they produce are the oilbird of South America, whose plump offspring used to be rendered down to make oil, and the edible-nest swiftlet of South-East Asia, whose nest is the key ingredient in the Chinese gastronomic delicacy, bird's nest soup. We might also allow the honeyguides, which do not actually produce honey, but do guide mammals (including us) to bees' nests, and the muttonbird, the colloquial name of the short-tailed shearwater of Australasia, whose harvested flesh is supposed to taste like mutton.
† The names cormorant and shag are, like pigeon and dove, or swallow and martin, more or less interchangeable.

Like two other seabird species found here – the Peruvian booby and the Peruvian pelican – the guanay cormorant takes full advantage of the Humboldt Current, which runs all the way along the Pacific coast and produces a constant upwelling of cold water, creating the ideal conditions for the vast shoals of fish (mostly anchovies and silversides) on which these birds feed. The cormorants catch their prey communally, congregating in huge flocks on the surface of the sea, before diving like synchronised swimmers to pursue their quarry underwater, reaching depths of up to 32 metres,[1] and occasionally as far as 74 metres beneath the waves.[2] The Humboldt Current, whose associated ecosystems have been described as 'the most productive marine environment on earth',[3] also creates very specific climatic conditions along these coasts: hot and dry, with virtually no rainfall. This, as we shall see, is the key to our story.

Like many seabirds, guanay cormorants nest colonially, choosing remote coastal headlands and offshore islands where they can be safe from land-based predators, and get easy access to their feeding grounds. The largest colonies are on clusters of rocky, volcanic islands just off the Peruvian coast, where the birds breed all year round, with a peak during the austral spring months of November and December.

The cormorants prefer to nest on flat surfaces or gentle slopes, on which the female lays two or three pale, bluish-green eggs, which she incubates for between three and four weeks. Scrawny and helpless when born and looking more like a fluffy reptile than a bird, the chicks are fed by both parents, fledging roughly four weeks after they hatch. Seabirds usually breed colonially because, unlike land-based birds, they do not need to defend their source of food, which is both abundant and easily accessible in the surrounding seas. So they nest in jam-packed, noisy and often very smelly colonies, cheek-by-jowl with their peers. Crowding together may lead to quarrels with the neighbours, but it also has the advantage of keeping the cormorants relatively safe from aerial predators. These include gulls, turkey vultures and South America's largest flying bird, the Andean condor, all of which patrol constantly overhead, hoping to snatch an unguarded egg or chick.

Each pair of guanay cormorants and their offspring produce copious amounts of droppings, which they then use to make their circular nest. Were the climate any wetter, the droppings would soon be washed away by rain, as happens in seabird colonies elsewhere in the world. But here, the almost totally arid conditions mean that, over the centuries, the guano has formed a thick crust on top of the rocky surface, up to 50 metres deep. It was this which, when harvested and sold for fertiliser, led to the species being dubbed 'the most valuable bird in the world'.

A major proportion of the riches generated by the guano trade were accrued by a nineteenth-century English businessman, William Gibbs, who would surely have appreciated that well-known saying, 'Where there's muck, there's brass.' Fuelled by the profits of his near monopoly of Peruvian guano – also known as 'brown gold'[4] – Gibbs eventually became the richest non-nobleman in England. He was celebrated (and maybe also mocked) in this popular verse:

> William Gibbs made his dibs,
> Selling the turds of foreign birds.[5]

Gibbs sank a large portion of his fortune into renovating Tyntesfield, a country estate to the west of Bristol.[6] But few of the 300,000 people who visit this splendid house and its gardens each year are aware of a darker aspect of its history. For the wealth that built Tyntesfield, and created the splendour we see now, came directly from the suffering of tens of thousands of those unnamed Chinese labourers, toiling in a living hell, described by one contemporary observer as 'a kind of human abattoir, or slaughterhouse of men'.[7]

In the longer term, the guano industry had an even more important influence on the modern world. It massively increased crop yields, which laid the basis for the way the land is still farmed today, and thus shaped the rural landscape of North America, Britain and much of the rest of Europe.

On a cold, but bright and sunny January afternoon, as rooks cawed in the distance, the grand house at Tyntesfield could hardly have appeared more bucolic and peaceful. It is the kind of project the National Trust does so well: a masterpiece of Victorian splendour, the house packed with more precious artefacts than any other Trust property, the lovingly tended arboretum and gardens complete with tea room and second-hand bookshop.

William Gibbs initially purchased Tyntesfield in 1843, for the sum of £21,295 (roughly £1.9 million at today's values).[8] The house had been rebuilt in the Georgian style in the early nineteenth century, on the site of a sixteenth-century hunting lodge. A decade later, using the proceeds of the guano trade, Gibbs commissioned the architect John Norton to improve the exterior of the house, doubling its size. He also hired the designer John Gregory Crace to remodel the interiors in the fashionable Victorian Gothic style, at eye-watering expense.[9] Over time, thanks to Gibbs's enormous wealth, Tyntesfield was transformed into what would become a unique example of this period, filled with thousands of paintings, books and works of art.[*]

My guide was the historian Dr Miranda Garrett, Property Curator at Tyntesfield. In the library, among more than 2,000 books, I noticed six immaculate volumes of the Revd Francis Orpen Morris's *A History of British Birds*. Along a corridor, we examined a stained-glass window decorated with what appear to be Peruvian boobies (another guano-producing bird), but may also show the guanay cormorants themselves. In pride of place at the top of the stairs was Sir William Boxall's imposing portrait of William Gibbs himself, at the age of seventy. It is a fine example of a Victorian gentleman who was, by then, secure in his wealth and consequent social status. Finally, we entered the specially built private chapel, glowing in the late afternoon light which illuminated the striking detail of the mosaics of Biblical scenes around the walls.[†]

It is easy to be beguiled by the sheer beauty of the house and its contents, their opulence and extravagance, as Miranda pointed out, surviving the passing of time. I was especially struck by the

---

[*] Tyntesfield remained in the Gibbs family for another three generations, until the death of his great-grandson George (known by his middle name of Richard), the second Baron Wraxall, in 2001. After a fundraising campaign, Tyntesfield was purchased for £24 million by the National Trust, as a unique example of the period.

[†] Gibbs began the chapel intending to be buried there, but died before it could be completed. Following a stand-off between his descendants and the Bishop of Bath and Wells, permission was refused, and he was interred in the nearby church.

hand-stencilled walls, the intricacy of the carved ceilings, and so many of the 74,000 precious artefacts.

The TV presenter Alan Titchmarsh was certainly bowled over: in the final episode of his Channel 5 series *Secrets of the National Trust*[10] he extolled Tyntesfield as 'one of the finest Victorian homes in all the land; [a] Gothic masterpiece . . . funded by fertiliser'. Unfortunately, those indentured workers, who mined the guano that funded William Gibbs's ostentatious lifestyle, were not mentioned even once. This is something Miranda Garrett is keen to change. The volunteer guides at Tyntesfield are well briefed on the darker history of the Gibbs family, and take pains to inform the visitors about how the money that funded the house and its contents was actually made.[11]

Today, if we are familiar with the word 'guano' at all, it is likely to be from pop culture. In the 1995 comedy *Ace Ventura 2: When Nature Calls*, a hapless detective played by Jim Carrey travels to Africa to investigate the trade in bat guano, worth billions of dollars.[12] And in Ian Fleming's novel *Dr No* (later turned into the first film in the James Bond franchise), Bond's evil adversary carries out his nuclear experiments on the fictitious island of Crab Key, off the coast of Jamaica. Crab Key is also a guano island – indeed, the eponymous villain funds his potentially apocalyptic work by mining and selling guano.[13]

But the story of guano is far older. The first recorded reference in the *Oxford English Dictionary* comes from the second year of the reign of King James I, in 1604, in a translation of a book originally written in Spanish by the Jesuit missionary José de Acosta, who writes of 'heapes of dung of sea-fowle . . . They call this dung Guano.'[14] Given that the Spanish conquistadores were searching for the 'three Gs – God, Gold and Glory', it is perhaps ironic that the greatest riches in the New World were to be found in another word beginning with 'G'; rather less refined, perhaps, but equally valuable.

Seabird guano is such a prized commodity because it contains very high proportions of three key elements – nitrogen, phosphate and potassium – along with smaller amounts of phosphorus, calcium and magnesium. All these are essential for the healthy growth of plants. Its apparently magical powers as a fertiliser were known by the indigenous peoples of the Pacific coast at least 1,500, and possibly as long as 5,000, years ago, well before it came to the attention of the wider world. We know this because, when layers of guano on the North and South Guanape Islands were removed during mining, hundreds of pre-Columbian objects were found. Dating back to early in the first millennium AD, they also included a grisly discovery: the bodies of young women, their naked bodies covered with thin gold foil, who had been brutally sacrificed.

Two other pre-Columbian cultures also exploited these guano reserves. The first was the Nazca civilisation, famous for the 'lines in the desert' that have inspired various outlandish theories about extra-terrestrial life, who lived along the southern coast of Peru from roughly 100 BC to AD 800. The other was the Inca Empire, the largest in the whole of the Americas before the arrival of the conquistadores, whose rule ran from the early thirteenth century until they were finally conquered by the Spanish in 1572. One popular science writer crudely – but accurately – described the Inca Empire as having been 'built on shit'.[15]

The place where the Incas lived was mountainous, arid and infertile. To grow crops successfully, they needed three things: flat strips of land, plenty of water, and a rich and effective fertiliser. To achieve this, they cut terraces to create level areas where the crops such as maize and potatoes could be planted, created complex irrigation systems to water them, and then used the guano collected from islands off the coast to provide the precious nutrients their crops needed to grow. This proved to be incredibly effective: producing up to three harvests a year on this previously harsh and uncultivatable land. Such was the success of this miraculous natural product that, among their many gods, the Incas worshipped the Huamantantac, whose name translates as 'He who causes the Cormorants to gather themselves together'.[16]

In his gripping history of the US guano trade, *Guano and the Opening of the Pacific World*,[17] Gregory T. Cushman cites the earliest written history of the Inca civilisation, written in 1609 by the Spanish-Peruvian chronicler Garcilaso de la Vega, known as 'El Inca'.[18] De la Vega documents the methods used to harvest the bird droppings, most of which came from the Chincha Islands, about 160km south of Lima. He explains that the right to take manure from each island was assigned to a particular province or provinces, and then allocated to the villagers according to their needs. Draconian laws were passed, records El Inca, to protect the birds from which the precious guano came:

In the times of the Inca kings these birds were so carefully watched that no one was allowed to land on the islands during the breeding season under pain of death, so they should not be disturbed or driven from their nests. It was also illegal to kill them at any season either on the islands or elsewhere, under pain of the same penalty.[19]

It has been suggested that the Incas' care for the seabird colonies proves that they were the world's first conservationists. However, the claims of several other candidates aside,* it seems likely that their primary motive for protecting the seabirds was not any altruistic concern for their welfare, but simply to safeguard their own interests.

Ironically, by the time de la Vega's work appeared, in the first decade of the seventeenth century, the Inca civilisation was no more. It had been first ravaged, then eradicated, by the European conquerors, who had brought smallpox and influenza to the Americas, against which the indigenous peoples had no defence.

Guano was first taken to Europe around the start of the eighteenth century, arriving in the southern Spanish seaport of Cádiz. Why the returning seafarers took the trouble to carry this malodorous substance back with them is something of a puzzle: there is no evidence that it was being used as a fertiliser outside South America, and many people commented adversely on its 'unfamiliar

---

* Far better claimants include the seventh-century monk St Cuthbert, who protected eider ducks on Lindisfarne, and the fourteenth-century Princess Eleonora of Arborea in Sardinia, who protected the falcon that still bears her name.

stench'. The breakthrough came during a voyage to the continent by one of the greatest scientists, explorers and polymaths of all time, Alexander von Humboldt – after whom an ocean current, a hummingbird and a South American penguin are all named.[*]

Humboldt became aware of guano in 1802 when he came across fields near Lima being fertilised with the substance. He brought back samples and sent them to Paris to be analysed by two leading French scientists, while another sample found its way into the hands of the pioneering English chemist Sir Humphry Davy. Davy was interested in the theories of the economist Thomas Malthus, whose influential 1798 work, *An Essay on the Principle of Population*,[20] had made an apocalyptic prediction: that human population growth, if left unchecked, would inevitably bring about widespread food shortages, famine and mass deaths. Davy responded to this challenge pragmatically: by launching investigations into how chemicals – especially nitrogen – might improve soils, and so increase crop yields, to produce more food. He had observed that Peruvian guano – with its very high concentrations of beneficial chemicals – had enabled 'the sterile plains of Peru' to become fertile. Clearly this merited further investigation, and by 1813, in his book *Elements of Agricultural Chemistry*, he was going as far as to suggest that God Himself had put 'the modification of the soil, and the application of manures . . . within the power of man'.[21]

However, the vast expense of transporting guano from South America to Europe, along with the readily available alternative of

[*] Humboldt's sapphire *Hylocharis humboldtii*, found in parts of Central and South America, and the Humboldt penguin *Spheniscus humboldti*, which feeds alongside the guanay cormorants off the coast of Peru.

'night soil' – human excrement collected from latrines – meant it was not at that stage economically viable to import it. Another major barrier was the length of time it took to ship the guano back: between three and eight months.

During the 1820s, after a series of hard-fought wars, and almost three centuries after the original Spanish conquest, Peru finally gained *de facto* independence from Spain, although this was not formally acknowledged for more than half a century. Peru's new government soon recognised the commercial value of seabird guano, and began to send larger quantities to Europe, where more trials were carried out to determine its efficacy as an agricultural fertiliser. The turning point came in 1838, when two Franco-Spanish merchants sent samples of Peruvian guano to a merchant and ship owner from Liverpool, William Myers. Myers, who was also a farmer, was impressed, and decided to take a gamble by investing in a much larger quantity of the product.[22]

On 23 July 1839, the ship *Heroine* finally docked at Liverpool after a long voyage from Valparaíso in Chile, carrying Myers' thirty bags of guano. Soon afterwards, he signed a contract with a Lima-based businessman, Don Francisco Quirós, to produce Peruvian guano specifically for the British market. The timing could hardly have been better. As a result of the economic and social changes wrought by the Industrial Revolution, the population of England and Wales would grow almost threefold in just over a century, from 6.1 million in 1750 to 17.9 million by 1851. All these people, many of whom now lived in towns and cities, needed feeding, and the countryside was beginning to suffer the consequences of this rapid rise in demand. It looked as if Malthus's dire predictions of starvation and famine might be about to come true. At about the same time, scientists were becoming increasingly concerned about the environmental costs of the drive for higher crop yields, and the resulting loss of crucial nutrients from the soil. The German scientist Justus von Liebig – known as 'the father of agricultural chemistry' – was warning against the over-exploitation of the land and emphasising the importance of nitrogen for healthy plant growth.[23]

Guano offered the perfect solution: it contained all the nutrients that plants require to grow, and so spectacularly outperformed any existing fertilisers. Once a farmer decided to use it, he would never want to go back to the old practices. One contemporary publication was quick to extol its efficacy on a global scale: 'The sunless lands of Great Britain, the rice fields of Italy, the vine lands of Germany, the exhausted coffee plantations of Brazil, and the arid plains of Peru, all testify to its fertilising properties.'[24]

The businessmen who had earned the right to import guano to Britain happened to be in the right place at exactly the right time, and their gamble paid off. Guano proved to be highly superior to the existing alternatives of night soil and horse manure, and crop yields skyrocketed. As did the revenues: the first large shipments earned roughly £100,000 (approximately £6.1 million at today's values).

Soon afterwards, another British businessman – William Gibbs – found himself taking a gamble on the guano trade. Ironically, William was initially very reluctant to enter into the contract that would ultimately make him rich beyond his dreams, calling it an 'act of insanity'. Nevertheless, against his natural inclination to be cautious, he was finally persuaded to go ahead.

William Gibbs may have been born over two centuries ago, in 1790, but his life and work suggest a close parallel with a contemporary William: Bill Gates. Like the Microsoft co-founder, having made a vast fortune through business, Gibbs became a generous philanthropist, in his case with a religious focus, through the restoration and building of churches.*

Again, like Gates, William Gibbs was a self-made man. Although he came from a respectable family, his father, a wool exporter, had at one time got into serious debt, potentially threatening his son's future in business before it had even begun. Those money worries forced the young William to leave school early, preventing him and his elder brother from attending university. Instead, at the age of sixteen, he was taken on as an office clerk at his uncle's firm in Bristol. Two years later, in 1808, he re-joined his father and brother at a new company of merchants in London.

In 1813, William became a partner in what was now Antony Gibbs and Sons and moved to the southern Spanish city of Cádiz. Over the next decade or so he worked hard to win new clients; and by the early 1820s had expanded the firm's business interests in South America, including Peru. The venture was a great success,

---

* William Gibbs's entry in the *Dictionary of National Biography* ends with a paean to his life and work, explaining that the monetary rewards generated by the guano trade 'enabled him to pursue the life of a model Christian gentleman and head of household. This was not a life of inward-looking piety, but a practical attempt to contribute towards the regeneration of Christian England.' No mention of the pain and suffering caused by the source of his wealth.

and William's life took another positive turn when, in August 1839, at the rather advanced age of forty-nine, he married the twenty-one-year-old Matilda Blanche Crawley-Boevey. Known as Blanche, she was, like her husband, an ardent High Church Anglican. They would go on to have seven children: four sons and three daughters – four of whom sadly died in their twenties, three from that terrible scourge of Victorian families, tuberculosis.

In 1842, three years after William and Blanche's marriage, his elder brother George Henry died unexpectedly, making William the sole partner in Antony Gibbs and Sons. That very year, a major upturn in his fortunes occurred when his representatives in Lima signed the contract that would make them the sole importers of guano into the United Kingdom. In return for advancing funds to the Peruvian government, the partnership would be entitled to the profits from sales of the fertiliser. In that first year, they imported just 182 tonnes of guano; by 1856, that had grown to 211,000 tonnes, peaking in 1862 at 435,000 tonnes.

The guano boom was well and truly under way. One aristocratic landowner, the Earl of Derby, who served as Britain's prime minister on no fewer than three separate occasions, was so delighted with the new product that he bought a whole shipload to use on his extensive Lancashire estates – not just to fertilise arable crops, but on fruit trees and bushes as well. Another farmer was so keen to obtain the precious substance that he attempted to steal it. Unfortunately, he did so by scooping it into a sack which he was holding open with his mouth. He accidentally swallowed some and died in agony the next day.

In less than four decades, from 1840 to 1879, an estimated 12.7

million tonnes of guano, worth between £100 million and £150 million (approximately £6.1 to £9.1 *billion* at today's values), was shipped from Peru to Europe and North America. For the Peruvian government, this provided a much-needed financial windfall, to the extent that by 1847 guano had become the country's most valuable export.[25]

By now, the word 'guano' had become so familiar that in the same year, in his novel *Tancred*, the future British prime minister Benjamin Disraeli used it as a witty metaphor for intellectual self-improvement: 'Lady Constance . . . having guanoed her mind by reading French novels, had a variety of conclusions on all social topics.'[26]

Given his initial antipathy towards the trade, it is ironical that William Gibbs benefitted more than anyone from the guano boom. From 1847, after falling out with his original partners, Gibbs acquired sole possession of the trade and the vast income it generated, which lasted until the early 1860s, when the company finally ended its association with the product.

In the light of subsequent events, it appears that the timing of Gibbs's exit was spot-on. By the late 1850s and early 1860s, new, cheaper and more easily available fertilisers had come onto the market, many of them targeted at a specific soil type or crop. These products, known as superphosphates, were both easily available and considerably less costly than the imported guano; they also worked far better on one of the staple crops, turnips. The die was cast: from now on, the kind of intensive agriculture first made possible by guano had to be met by the development of increasingly effective fertilisers.

Back in Peru, two further problems accelerated the decline of the guano trade. The first was a dispute over ownership, leading to the 'First Guano War', fought by Spain against Peru and Chile, which began in April 1864 when Spanish troops occupied the Chincha Islands in order to seize their share of this lucrative global trade, and reassert sovereignty over their former colony. The occupation continued for two years, during which exports collapsed, and Peru's economy – by then heavily dependent on revenue from the sales of guano – suffered a major recession.

By then, the second – and far more serious – issue had arisen. Greed had led to a massive over-exploitation of this theoretically renewable, but in practice limited, resource. By the 1870s, the deposits on the Chincha Islands – the main source of the guano exported to Europe – had effectively run out, as the birds simply couldn't produce enough droppings to replenish what had been so rapidly taken. The goose (or in this case the cormorant) that laid the golden egg was no more. Alexander Duffield was an English mining engineer who visited the islands both before and after the mass harvesting of guano. Beforehand, he observed, they were 'bold, brown heads, tall and erect, standing out of the sea like living things, reflecting the light of heaven . . . Now these same islands looked like creatures whose heads had been cut off . . . like anything in short that reminds one of death and the grave.'[27]

The short-lived Peruvian guano boom – the South American equivalent of the California Gold Rush – was over. However, its consequences live on, in two ways: the changes it kick-started on the way the land is farmed in North America and Europe, and a shameful legacy of human suffering.

*

The Chinese workers who mined the guano are often referred to as slaves; and in practice, that's exactly what they were. As one contemporary observer noted, they were 'without any title, or rights . . . mere over-worked beasts of burden'.[28] But their contracts stated that they were not actually slaves, but indentured workers, which meant that they were trapped in a terrible limbo; never able to earn enough to pay back the costs of their voyage and their meagre rations of food, and so with no realistic prospect of attaining freedom. Some had not even signed on voluntarily, having been press-ganged – kidnapped against their will – and forced onto the waiting ships.

Alexander Duffield witnessed the appalling treatment of the Chinese labourers for himself. 'No hell has ever been conceived . . . that can be equalled in the fierceness of its heat, the horror of its stink, and the damnation of those compelled to labour there, to a deposit of Peruvian guano when being shovelled into ships.'[29] Another English engineer, George Fitzroy-Cole, observed that 'their lot in these dreary spots is a most unhappy one.' Writing in 1870, the United States Consul, D. J. Williamson, described a way of life so terrible that many of the workers regarded death as a merciful release. To prevent them, 'in moments of despair', from committing suicide by throwing themselves into the sea to drown, he noted, guards were posted along the edges of the cliffs around each island. Their miserable lives were summed up by the US historian Watt Stewart, from whose seminal 1951 work *Chinese Bondage in Peru*[30] these harrowing accounts come:

The condition of the Chinese coolie in Peru was lamentable. In no case was he brought to Peru for his own betterment.

He was there to serve the interests of the Peruvian master . . .
He was scarcely regarded as a human being – rather he was
a machine for the production of wealth. His ills – physical,
social, and psychological – were indeed appalling.

The statistics on the workers' chances of survival are truly horrific:
during the first fifteen years of the industry, the annual mortality
rate was between 35 per cent and 40 per cent. Few managed to
survive even to the nominal end of their five-year contract.

In 1854, the Peruvian government abolished slavery, but for the
Chinese labourers in the guano industry this made virtually no
difference. They were still stuck in a classic Catch-22: unlike those
people brought to Peru from Africa, their status as indentured
workers meant they were not eligible to gain their freedom because
they had never been legally defined as slaves. Abolition might even
have made things worse: now that African slaves were no longer
available, Chinese labourers were in even greater demand.[31] And
when, in 2009, Peru's government finally made a formal apology
for the country's past abuse of immigrants, this was directed solely
at the Peruvian descendants of African slaves. The Chinese guano
workers were never mentioned.

It is easy to forget that at the heart of this tale is a real, wild bird.
How did the guano trade affect the welfare, numbers and long-
term status of the guanay cormorant itself?

We know that the daily disturbance by the labourers, added to
the longer-term damage to the cormorants' breeding habitat, and
the taking of eggs and birds for food, led to the total destruction

of some colonies, and a major decline in others. So, it would be reasonable to assume that the guano industry was bad for the seabirds. One estimate suggests that numbers plummeted by well over 90 per cent, from an estimated 53 million birds in the late nineteenth century, to just 4.2 million by 2011.

Things could actually have been a lot worse were it not for actions taken by the Peruvian authorities to safeguard the birds. From the early twentieth century onwards, long after the peak of the guano trade was over, they designated the islands as some of the world's first nature reserves, and then managed the cormorants as if they were domesticated livestock.* Measures included allowing the birds to finish nesting before the harvest began each year; posting guards to prevent intruders from killing them or taking their eggs; and allowing breaks in the harvesting so that the guano reserves could recover.[32] Today, BirdLife International categorises the guanay cormorant as 'near threatened', because it has 'experienced moderately rapid declines in the past three generations (thirty-three years)'.[33]

One of the biggest threats facing all seabirds along the Pacific coast of South America is being accidentally caught in nets intended to catch fish; especially anchovies, on which the cormorants feed. In northern Peru, as many as 20,000 cormorants are also trapped or shot each year as food for humans.[34] [35] In the longer term, the

* Although this did help reverse the fall in guanay cormorant numbers, it unintentionally led to the decline of several other seabird species, including the Humboldt penguin, which is now considered as 'vulnerable', with fewer than 24,000 birds remaining in the wild. The cormorant's largest predator, the Andean condor, was also shot mercilessly by the seabirds' guards. It too is now categorised as 'vulnerable', with just 6,700 individuals remaining.

climate emergency poses a major issue for the guanay cormorant and many other Pacific Ocean seabird species. Global heating is almost certainly the cause of the recent rise in the frequency of 'El Niño' events, during which ocean temperatures in the eastern Pacific become far warmer than usual, dramatically disrupting the seabirds' food supplies. The 1982–3 El Niño led to a disastrous breeding season, and the deaths of as many as 1.7 million guanay cormorants – close to half the entire world population; since then, however, numbers have managed partly to recover.[36]

Meanwhile, guano is still being harvested and exported, though the annual yield in the tens of thousands of tonnes is far lower than the mid-nineteenth-century peak, when almost half a million tonnes was produced each year.[37] Nevertheless, it has been estimated that globally, guano's worth to the world economy may still be as much as one *billion* US dollars a year.[38]

It is, perhaps, easy to regard the whole guano industry as a rather curious historical episode, with little or no relevance to the modern world. But you only need to compare the grounds of Tyntesfield (and the surrounding sustainably farmed land) with the rest of England's lowland countryside, to see the legacy of that first guano-fuelled boom in intensive farming. Where the fields in and around Tyntesfield are rich and varied, and echo to the buzz of insects and the sound of birdsong (because they are managed sustainably by the National Trust), those elsewhere are mainly sown with rapidly growing ryegrass, grown either for grazing sheep and cattle, or to be cut for silage, to feed intensively reared livestock for milk or meat.

Farming may still retain its homely image, with the industry assiduously promoting farmers as the 'custodians of the countryside', but the reality is very different. Farming in lowland Britain, North America and elsewhere in the developed world is an industry, pure and simple. The dominant model is a high-input, high-output system, using vast amounts of chemical fertilisers, herbicides and pesticides to produce the greatest possible yields of food, at the lowest possible cost for retailers and consumers. This is reflected in the exponential growth of the production of phosphate- and nitrogen-based fertilisers during the twentieth century: the former growing by a factor of more than 40-fold; the latter by over 250-fold.[39]

The disastrous consequences for biodiversity of this industrial approach to agriculture, relying on chemicals, are now abundantly clear. The wildlife that used to thrive in the countryside has been pushed to the margins, with many species of birds, mammals, insects and wild flowers either eradicated as 'pests', or their numbers reduced to a fraction of what would once have lived there. This is largely due to a fundamental change in attitudes and approach to agriculture, which has its roots in the use of guano as a fertiliser, more than a century ago.

Looking back, we can see that almost from the very beginnings of the guano industry, the seeds of its decline were already being sown. The product was expensive, difficult to transport, and unreliable in terms of delivery. It was also, ultimately, finite, as supplies began to run out and new sources obtained elsewhere proved to be less effective and not commercially viable. Yet, even though the guano boom may have lasted for little more than thirty years – less than a single human generation – it changed the working lives of farmers for ever, by opening their minds to the prospect of what was then called 'high farming' – a forerunner of the now-dominant industrial farming system that has so blighted the modern world.[40]

Guano's key influence was, paradoxically, in hastening its own demise: kick-starting the search for an alternative, cleaner and more convenient substance. The aim was to reduce the northern hemisphere farmers' dependence on this messy and unreliable product, from the other side of the world, and replace it with one that was cheap, consistent and easily produced: artificial fertilisers. As Gregory Cushman points out, that quest was a direct result of the discovery of huge supplies of natural fertiliser – in the form of guano – during the second half of the nineteenth and early part of the twentieth centuries.[41]

For decades, scientists had struggled to find a method of producing ammonia by fixing nitrogen on an industrial scale, with several processes developed, but abandoned as being too costly and inefficient. Then, during the opening decade of the twentieth century, two German chemists, Fritz Haber and Carl Bosch, achieved a crucial breakthrough, in what became known as the Haber-Bosch process – still the main way of producing ammonia today.[42]

In 1918 and 1931 respectively, Haber and Bosch were awarded the Nobel Prize for Chemistry for their achievements.[*] Their

---

[*] Haber's award was highly controversial: his process may, in the words of the committee, have given 'the greatest benefit to mankind', but it had also been used to produce the chemical weapons deployed by the German military to such devastating effect during the First World War. At the awards ceremony, the eminent physicist Ernest Rutherford (who had himself been awarded the prize a decade earlier) refused to shake Haber's hand.

breakthrough had opened the floodgates to the mass production of artificial fertilisers.[43] This transformed both the farming industry and the rural landscape across much of the northern hemisphere and, as an unintended consequence, pushed wildlife to the periphery. While this shift began during the early 1900s, it accelerated dramatically because of the urgent need for food during the twentieth century's greatest conflict, the Second World War.

It would be an exaggeration, albeit only a minor one, to say that the British did not see the coming of the Second World War until it was too late. Yet the unpalatable truth is that a combination of complacency and an outdated belief in the place of honesty in geo-politics saw the threat from Adolf Hitler and Nazi Germany grossly underestimated during the lead-up to the conflict. Consequently, the country was woefully underprepared for the immediate outcomes of the declaration of war, not just in terms of armaments, but also something far more basic: food. Put simply, 41 million British men, women and children were under serious threat of starvation.

At the outbreak of war in September 1939, the government hurriedly resurrected the War Agricultural Executive Committees (originally created during the First World War, but then allowed to lapse) for every county in England and Wales. Their urgent mission was to increase food production, and they were given wide-ranging powers to do so, as one contemporary report from the committee reveals.[44] The immediate aim was to increase the area of arable land under cultivation by at least 600,000 hectares by the time of the following year's harvest, a mind-bogglingly difficult task. Already overworked farmers had little or no choice but to co-operate: if they refused, their land could be confiscated.

This combination of an appeal to farmers' patriotism and the threat of sanctions worked: by spring 1940, less than a year after the measures were introduced, the total area of productive land had actually risen by almost 700,000 hectares – about the same size as Lincolnshire or Devon. A *Pathé Gazette* newsreel film from that time, entitled 'Food from Waste Land', strikes a triumphant note when documenting the undoubted success of this policy: 'A short while ago, this was the 6,000-acre wilderness of Feltwell Fen, in south-west Norfolk, where nothing grew, save reeds and weeds . . . But it has taken a war to turn that same wasteland into an agricultural goldmine. The Ministry of Agriculture has set to work an army of men reclaiming the idle acres.'[45]

From today's perspective, this may look like typical – and unintentionally amusing – wartime propaganda. But with the very real need to increase food production in the face of German naval blockades, it was clearly necessary. Unfortunately, as the nature writer Mark Cocker points out, the so-called 'wasteland' that disappeared beneath the plough included many nature-rich habitats such as ancient woodland, scrub, heaths and hay meadows.[46] This might not have proved so disastrous were it not for the fact that even after the end of the conflict, and the eventual with-drawal of food rationing in the early 1950s, the same mindset that had so massively increased food production during wartime continued to be applied to post-war Britain. This went hand in hand with the introduction of government (and later European Union) subsidies, and the rise in power and influence of the companies that manufactured the chemicals required.[47]

As Dr Rob Lambert of the University of Nottingham explains, at the time this approach was seen as the gateway to a bright new future, for both the countryside and Britain as a whole. The aim, from the late 1950s and well into the 1960s, was to get rid of what were seen as outdated and inefficient pre-war farming methods and replace them with cutting-edge new technologies. One of them was the widespread use of pesticides, and the birth of what we now know as 'chemical farming'.[48]

What nobody had foreseen were the consequences of this total dependence on chemicals. For by exterminating weeds, insects and other 'undesirables', the system was also hastening the decline of the more benign flora and fauna of the British countryside: skylarks and grey partridges, beetles and butterflies, cornflowers and cowslips, and many other once common species, went into a rapid decline.

The same was happening throughout the developed world, especially in North America, where chemical farming had been embraced with even greater enthusiasm, and the product DDT (Dichlorodiphenyltrichloroethane) had been hailed as a miracle insecticide. As a result, common songbirds such as the bobolink, and raptors higher up the food chain including the peregrine, saw their numbers go into freefall. For the bald eagle, the very symbol of the USA, this almost resulted in the global extinction of the species (see chapter 8).

In 1962, the US environmentalist Rachel Carson published *Silent Spring*,[49] the first book to reveal the true environmental consequences of the chemical revolution in farming. Even so, the use of DDT, the best-known of the many chemicals that caused a catastrophic decline in farmland birds, was not actually banned in the United States until a decade later, in 1972,[50] and not finally withdrawn in the United Kingdom until the mid-1980s.[51] In Latin America, tens of millions of overwintering songbirds of over 150 species, which then travel north to breed in the USA and Canada, are still being wiped out by the widespread use of toxic pesticides. Paradoxically, these pesticides are mainly used to grow out-of-season fruit and vegetables, to meet the demand from consumers in the lucrative North American market.[52]

But the greatest warning sign comes from plummeting insect populations on both sides of the Atlantic, memorably described by Professor Dave Goulson of the University of Sussex as 'ecological Armageddon'.[53] This is due in large part to the development and widespread use of a new type of insecticide, neonicotinoids, which not only kill harmful 'pests', but also appear to be fatal to more benevolent – and useful – insects and birds. It could be – indeed has been – argued that a second 'Silent Spring' has already begun.[54]

This was confirmed in 2019, when two ecologists, Francisco Sánchez-Bayo and Kris A. G. Wyckhuys, published a meta-study reviewing evidence from over seventy papers documenting insect declines in Europe, North America and elsewhere.[55] Their conclusions were stark: insects are declining at a rate eight times faster than vertebrate species such as mammals and birds, with over 40 per cent of all species threatened with extinction. 'We are witnessing the largest extinction event on Earth,' the authors concluded, 'since the late Permian' (a geological epoch 250 million years ago).[56]

There is a growing body of evidence proving that insects are crucial to global economic success – especially, ironically, for the agricultural industry. In 2006, two entomologists, John Losey of Cornell University and Mace Vaughan of the Xerces Society for Invertebrate Conservation, worked out the value of pollinating and predatory insects to the US economy: a cool $57 billion a year – close to half the entire annual contribution to that economy from all America's farms put together.[57]

Yet despite a growing movement that challenges the current status quo, and offers sustainable alternatives, farming throughout the developed (and much of the developing) world is still largely based on the same high-input, high-output system initially tried out back in the mid-nineteenth century when guano became the first widely available fertiliser. As Justus von Liebig foresaw as early as 1859, too many people continue to assume that the soil is an infinite and inexhaustible resource and fail to see that fertilisers are simply a short-term 'sticking plaster', which fails to address the long-term problem of soil degradation.[58]

The ultimate irony is that, because guano was so effective as a fertiliser, it masked the essential problem with all industrial farming methods: that by artificially prolonging the life of the soil, farmers fail to create a sustainable, long-term and wildlife-friendly method of producing food. It also changed the world, even if most people remain unaware of its importance, as Gregory T. Cushman wryly points out. 'No one needs to be convinced that the Black Death, African slave trade or Second World War fundamentally altered the course of human development. It is quite another thing to convince you that guano is of comparable importance.'[59]

But it is. And this – not the magnificent buildings and grounds of Tyntesfield – is the true legacy of William Gibbs.

7

SNOWY EGRET

Egretta thula

A bird in the bush is worth two in the hand.

Slogan of *Bird-Lore*, magazine of the
Audubon Societies of North America, 1899

*The eighth of July 1905 was just another hot and humid summer's day
in the Florida Everglades, towards the southernmost tip of the mainland United States. But although he did not know it, for Guy Bradley,
it would be his last day on earth.*

*Bradley was described by friends as 'pleasant, quiet, fair, with
blue eyes . . . clean-cut, reliable, courageous, energetic and con-
scientious'.[1] Yet he had not always been on the side of the birds. As
a youth, he had accompanied the notorious French plume hunter
Jean Chevalier on a collecting expedition through southern Florida,
during which the group had shot almost 1,400 birds of thirty-six
different species.*

*By the turn of the century, though, Bradley had changed his mind,
choosing to protect birds rather than kill them. A classic 'poacher
turned gamekeeper', he went on to become one of America's first official
wildlife wardens.[2]*

*So, when he heard gunshots that morning near his home on the
waterfront, he immediately went to investigate. Close by, he came
across three men: American Civil War veteran Walter Smith and his
two adult sons. As Bradley expected, he had caught them red-handed,
just as they were loading their booty onto a boat. Their grisly haul
comprised dozens of dead waterbirds, whose feathers, destined for the
fashion trade, would yield a substantial sum of money.*

Given the lack of eyewitnesses, we cannot be sure exactly what hap-pened next. It appears that Bradley confronted the trio and told them he would be arresting them for breaking the law. Then, just as he stepped forward to do so, Walter Smith raised his hunting rifle, and shot Bradley in the chest at point-blank range.

When Bradley failed to return home that day, his wife alerted the authorities, and a search party was sent out. The following day, his older brother Louis found the body, which had either fallen or been dumped in the creek and drifted some distance from the crime scene. Guy Morrell Bradley – wildlife warden, husband and devoted father – had bled to death from his wounds, aged just thirty-five.[3]

As with many 'true crimes', unravelling what really happened that day is a complex, messy affair. But there's one thing we do know: Guy Bradley and Walter Smith had a long history of mutual antag-onism. Bradley had arrested Smith and his oldest son on several previous occasions,[4] following which Smith had issued a chilling – and, as it turned out, prophetic – threat: 'If you ever arrest one of my boys again, I'll kill you.'[5]

Realising that if he now went on the run, this outburst would surely count against him, Walter Smith sailed over to Key West and turned himself in to the authorities. But this was not the remorseful gesture it might first appear. At his trial, Smith claimed self-defence, telling the jury that Bradley had fired first, but missed his target. Because the only eyewitnesses were Smith's own sons, the prosecution could not disprove this patently false testimony. Smith was found not guilty of murder and walked free from the courtroom.

It turned out to be a Pyrrhic victory. Smith's home was burned down in revenge by Bradley's brothers-in-law, while Bradley's widow Fronie and their two young children were given a new home in Key West, thanks to public donations collected by a recently founded environmental organisation, the Florida Audubon Society.[6] Bradley's murder also sparked a much wider public outcry, with newspaper headlines expressing outrage at the crime, not just in his home state, but all over the United States.[7]

One month after his death, Guy Bradley's obituary in the August 1905 edition of the Audubon Societies' magazine *Bird-Lore* described him as

A faithful and devoted warden . . . cut off in a moment, for what? That a few more plume birds might be secured to adorn heartless women's bonnets. Heretofore the price has been the life of the birds, now is added human blood. Every great movement must have its martyrs, and Guy M. Bradley is the first martyr in bird protection.[8]

Guy Bradley may have been the first, but he was certainly not the last, in the long, bloody and still continuing battle to protect the world's birds. In November 1908, just three years after Bradley's murder, Columbus G. McLeod, a game warden and deputy sheriff of DeSoto County north of Fort Myers, was reported missing. A month later, his boat and body were found; McLeod had been violently killed by axe blows to his head. That same year, Pressley Reeves, who was working for the South Carolina Audubon Society, was, like Bradley, shot and killed. The perpetrators of these murders were never brought to justice.[9]

Yet despite – or perhaps because of – the lack of payback for these terrible crimes, the tide would finally turn against the plume hunters, and towards the conservationists; away from killing and towards protecting birds. During the next two decades, a cruel and destructive industry, which at its height was worth tens of millions of dollars, would finally, like both the human and avian victims of its unthinking cruelty, die an undignified death.

To understand the background to those callous murders, and the corresponding rise of wildlife conservation organisations such as the Audubon Societies, we must tell the story of a slim and stylish species of heron that, at the turn of the twentieth century, found itself at the centre of a cruel, destructive and very profitable industry: the plumage trade.

The snowy egret – the New World equivalent of the Old World's little egret – is a small member of the heron family, found across a broad swathe of the Americas, from Nova Scotia in the north to Patagonia in the south.[10] As its name suggests, it has a dazzling white plumage, and a strikingly graceful and elegant appearance.

Not surprisingly, this beautiful creature caught the eye of the doyen of nineteenth-century bird artists, John James Audubon, who painted the snowy egret in his mammoth work *Birds of America*, published from 1827 to 1838.[11] Like all artists of his time, in the absence of optical aids, Audubon would have used a stuffed and mounted specimen of the egret as his model.

At the time, killing birds for art or science was not regarded as presenting any kind of ethical contradiction or conservation concern. After all, species such as the snowy egret were so incredibly abundant that the idea that they might become endangered was simply inconceivable. Audubon had himself visited colonies where thousands of pairs of egrets were nesting, writing that they were so abundant that, when they took to the air, their flocks would momentarily block out the light from the sun.

Audubon's painting shows a handsome creature, depicted virtually life-size, in full breeding plumage: snowy-white, with a sharp black bill and patches of yellow in front of its eyes, as it stands amongst aquatic vegetation, ready to pounce on a fish or a frog.[12] But the most striking feature of this remarkably life-like portrait is the set of long, feathery plumes extending from its crown, chest and wings, which are used by both male and female egrets for sexual advertisement in their courtship displays. These clearly mesmerised Audubon: 'Every now and then they utter a rough guttural sort of sigh,' he wrote, 'raising at the same moment their beautiful crest and loose recurved plumes, curving the neck, and rising on their legs to their full height, as if about to strut on the branches.'[13]

Ironically, although those plumes were crucial to the egrets' breeding success, they would almost prove to be their downfall. For those particular feathers – known as 'aigrettes' – were in great demand from the fashion trade on both sides of the Atlantic, where milliners would use them to adorn the hats of rich and beautiful high-society women.

These women's vanity fuelled the huge profits to be made by

rich traders, who would pay poor men like Walter Smith a small sum to collect the feathers, and then sell them on to the industry at inflated prices. At the height of the plumage trade, in the early 1900s, snowy egret feathers might sell for as much as $32 an ounce: equivalent to about $860 at today's values,[14] and in those days more than the price of gold.[15] This apparently insatiable demand would end up driving the snowy egret close to extinction.

The egrets' plight was worsened by their habit of breeding in large, noisy and very obvious colonies, known as 'rookeries', which made it easy to kill and collect them in huge numbers. The method used by the poachers was simple but effective: approach the rookery by boat and, once within range, simply shoot every adult egret they could see. Bringing them aboard, they would then skin the birds or pluck the feathers from the still warm corpses, throwing the carcasses overboard to rot in the water. If there were eggs left in the nest, these would soon be taken by predators such as turkey vultures or crows; any chicks would be predated, starve to death or perish in the baking summer sun.

On a visit to one colony, in the aftermath of a plume raid, the conservationist Gilbert Pearson was appalled at what he saw, subsequently writing of his horror at seeing the corpses of adult egrets with 'the skins bearing their plumes stripped from their backs', and the even more distressing sight of 'young orphan birds . . . clamouring piteously for food which their dead parents could never again bring them'.[16] Even the poachers were not immune to the consequences of their actions; as one admitted: 'The heads and necks of the young birds were hanging out of the nests by the hundreds. I am done with bird hunting for ever!'[17]

When Guy Bradley applied for the job as wildlife warden, in 1902, he wrote to William Dutcher, President of the Florida Audubon Society, confessing to his past mistakes, and recanting his time as a hunter: 'I used to hunt plume birds, but since the game laws were passed, I have not killed a plume bird. For it is a cruel and hard calling not withstanding being unlawful. I make this statement upon honour.'[18]

Though some former hunters did, like Bradley, admit the error of their ways, the rewards for those who chose to keep pursuing the birds were hard to resist. In 1896 one leading hunter, David 'Egret' Bennett, confessed all in a lurid interview with the New York *Sun* newspaper, freely admitting he had driven the snowy egret to the edge of extinction in North America, while revelling in the personal fortune he had made.[19]

Meanwhile, the cause of those who wished to see hunting stopped was not helped by the intransigence of some professional ornithologists and collectors, who claimed to see nothing wrong in the killing of birds. They may also have had an ulterior motive, fearing that if the plume trade were banned, they would no longer be able to shoot specimens to add to their museum collections. As Charles B. Cory, president-elect of the American Ornithologists' Union famously responded, when asked about the need for bird protection laws: 'I do not protect birds. I kill them.'[20]

Long before then, in the mid-1880s, the American Ornithologists' Union had estimated that as many as five *million* waterbirds were being killed every single year. At that rate, even species as abundant as the snowy egret would go extinct within decades, if not sooner. Already, there were no remaining colonies near either of Florida's largest cities, Tampa and Miami, and even those in the more remote parts of the Everglades were at serious risk of disappearing. By the early 1900s, egret colonies in Florida had been so depleted that plumes had to be imported from South and Central America instead.

But before we find out how this particular species and its relatives were ultimately saved from extinction, we must discover how feathers became such a valuable and sought-after resource in the first place. This goes back more than a century before the height of the plume trade, to one of the most notorious figures in European history: Marie Antoinette.

From long before dawn on 16 October 1793, crowds began to gather in Paris's Place de la Révolution, to witness the execution of a queen. And not just any queen, but Maria Antonia Josepha Johanna, better known as Marie Antoinette; the wife – and now widow – of the last King of France, Louis XVI. Nine months earlier, Louis had been executed by the guillotine; now, on this chilly autumn morning, it was her turn.

Marie Antoinette was – still is – a byword for conspicuous consumption and needless excess. That, when told the French people were so poor they could not even afford to buy bread, she said, 'Let them eat cake,' is probably a myth, spread by her revolutionary opponents to discredit her.[21] But one thing we do know for certain: she dressed with remarkable extravagance.

In one of more than thirty portraits of Marie Antoinette by the court artist Madame Élisabeth Vigée Le Brun, the queen is shown with her hair piled high on her head, wearing a hat decorated with ostrich feathers.[22] When, a few years later, Madame Le Brun painted her own self-portrait, she too sported a prominent feather in her hat.[23] When Marie Antoinette was led to the scaffold her hair had been crudely cut, and her head covered with a simple cap – visually symbolising her dramatic fall from grace.

Before the 1789 revolution, the queen's preference for wearing feathers in her hats had become a popular affectation among upper-class women in France and beyond. The feathers themselves were mainly from large and showy birds such as ostriches, peacocks, pheasants and storks, rather than herons and egrets, but they launched a global trade.

As often happens in the world of high fashion, what began as a whim of the rich soon filtered down to the middle classes – the bourgeoisie of North America and Europe – who flocked to department stores in London, Paris and New York to buy the very latest designs. At the height of the demand for feather-based creations, there were no fewer than 425 *plumassiers* (feather-dealers) in Paris alone, while across the Atlantic in New York, the millinery industry employed 83,000 workers, mostly working long hours in sweatshops, for very low wages.[24]

The combination of rising public demand and the money to be made meant that the quantity of skins and feathers required grew exponentially. During just three months in early 1885, 750,000 skins of snowy and little egrets were sold at auction houses in London alone. During the first decade of the twentieth century, over 14 million lbs (more than 6 million kilos) of feathers, worth almost £20 million (well over £2 *billion* at today's values) were brought into the United Kingdom.[25]

It was not just colourful and exotic foreign species that were being targeted. This high level of demand inevitably led to the wholesale trapping and killing of native birds, too. One study of the plume trade listed the various different ways in which birds were killed: by being 'trapped . . . limed, shot, bludgeoned or poisoned by the thousands of underemployed rural poor who provided raw materials for the milliner.'[26]

Specially chartered trains, carrying hordes of 'beer-swilling plumage hunters', travelled from London to seabird colonies on the Yorkshire coast and the Isle of Wight. On the Yorkshire cliffs, they shot and trapped thousands of kittiwakes: delicate and rather attractive gulls, whose dove-grey, black-tipped wings were in great demand. The concept of cruelty did not even cross their minds; as one horrified observer noted, the hunters were 'cutting their wings off and flinging the victim into the sea, to struggle with feet and head until death came slowly to their relief'.[27]

Another bird ruthlessly targeted by the trade was the great crested grebe. Today, this elegant waterbird is a familiar sight on rivers, lakes and gravel pits throughout lowland Britain. Like other members of its family, it leads an almost entirely aquatic lifestyle: swimming, diving and even building a floating nest.

The great crested grebe has adapted to its watery world by evolving incredibly dense feathering on its breast and belly, to conserve heat. But, just like the egret's showy plumes, this adaptation would almost prove to be the great crested grebe's downfall. Those dense feathers, as soft as animal fur, were made into hand muffs, while the bird's other prominent plumage feature, its black and chestnut crests, known as 'tippets' by the fashion trade, were in great demand to decorate women's hats.

As a result, by 1860, the UK breeding population of great crested grebes had fallen to between thirty-two and seventy-two pairs, putting it at imminent risk of extinction as a British breeding bird.[28] Ultimately, new bird protection laws came into effect just in time to save the species.

Although London was the main centre of the trade, this was a truly worldwide industry. In his 1913 book *Our Vanishing Wild Life*, the American zoologist (and former taxidermist) William T. Hornaday outlined in stark terms the worldwide impacts of the fashion trade's insatiable demand for feathers:

> From the trackless jungles of New Guinea, round the world both ways to the snow-capped peaks of the Andes, no unpro-tected bird is safe. The humming-birds of Brazil, the egrets of the world at large, the rare birds of paradise, the toucan, the eagle, the condor and the emu, all are being exterminated to swell the annual profits of the millinery trade.[29]

The naturalist Malcom Smith notes that the demand for plumes ultimately resulted in the extinction of the unique Middle Eastern race of the ostrich.

Even after the fall and execution of Marie Antoinette, the killing continued, with more than 500 kilos of ostrich feathers imported into France in a single year, 1807. Smith suggests that such was the constant demand for ostrich feathers from the fashion trade that, had it not been for the creation of ostrich farms in South Africa from the 1860s onwards, the world's largest species of bird might have gone globally extinct.[30] He also calculates that, during the half-century from 1870 to 1920, more than 18,000 tonnes of bird skins and feathers were imported into the United Kingdom – equivalent to as many ten *billion* birds.[31] To put this into perspective, estimates suggest that there are currently about 50 billion birds on Earth, so the plume trade was clearly making significant – and for some species unsustainable – inroads into the overall global population.[32]

And it was not just large, exotic birds that were being targeted.

Today, Frank Chapman is one of the most celebrated figures in the history of North American birding. As well as his role as the founder of *Bird-Lore* (now the *Audubon Magazine*), he is best-known for the annual Christmas Bird Count (CBC), the longest-running citizen science survey in the world. Today, the Christmas Bird Count is a Pan-American event, with tens of thousands of volunteer birders counting not just every species they can see, but also how many individuals of each species, across all fifty US states, as well as in parts of Central and South America. But its origins are far more modest.

On Christmas Day, 1900, Frank Chapman and twenty-six fellow observers took part in the very first CBC, at twenty-five different sites across the United States and Canada. Once the results were collated, the final tally was 18,500 birds of ninety different species. Since then, the totals and scope of the count have grown, so that in 2018 – the 119th annual event – a record 79,425 participants counted almost 49 *million* birds of over 2,600 species.[33]

Frank Chapman would have been stunned – and no doubt gratified – at how huge the CBC has become. But perhaps even more important than the scientific data it produces is that it also symbolises the journey made by bird conservation since the turn of the twentieth century. For not only did Chapman base his idea on the earlier tradition of festive season 'side-hunts', where the birds were shot instead of counted, but – just like Guy Bradley – he was a reformed hunter himself. Notoriously, before he saw the error of his ways, he had shot no fewer than fifteen Carolina parakeets, North America's only native species of parrot, which, only a few decades later, went globally extinct.[34] Yet even when he was still shooting and collecting birds, Chapman was gradually becoming aware of the dangers to bird populations posed by the plume trade.

One summer's day in 1886, Frank Chapman took a stroll along what was known as 'Ladies' Mile' in Manhattan – the most fashionable place to shop not just in New York, but arguably in the world. When he realised that many of the women who walked past him were wearing ornate hats, each one decorated with bird skins and feathers, he decided to count all the species of bird he could see. Of the 700 hats he noted on this and a subsequent walk, roughly three-quarters sported feathers – from no fewer than forty identifiable species.[35] As the historian Douglas Brinkley records, millions of birds were being killed each year to meet the constant and growing demand for bird-related headgear, which was in turn becoming more and more grotesque: 'Some women even wanted a stuffed owl head on their bonnets and a full hummingbird wrapped in bejewelled vegetation as a brooch.'[36]

What was even more shocking to Chapman was the variety of birds involved. He saw not just waterbirds and gamebirds, as he had expected, but owls and woodpeckers, orioles and flycatchers, blue jays and bluebirds, swallows and sparrows, tanagers and terns, waxwings and warblers – each one shot, plucked, stuffed and mounted as adornments to the wearer's vanity. 'It is probable that few if any of the women', he remarked afterwards, 'knew that they were wearing the plumage of the birds of our gardens, orchards and forests'.[37]

Frank Chapman's outrage changed him from poacher to game-keeper, and helped kick-start the movement to protect America's birds which would ultimately bring an end to the plume trade. But in a further twist, it was not men but women who would lay the foundations for the industry's ultimate demise. They would launch the two organisations that fought against this needless annihilation of birds, and which remain at the centre of today's global bird pro-tection and conservation movement: the Audubon Society in the United States of America, and the Royal Society for the Protection of Birds in the United Kingdom.

In Britain, bird protection had actually begun at the very start of the plumage trade, with the passing of the 1869 Sea Birds Preservation Act. This was the first bird protection legislation to be passed, not just in the United Kingdom, but anywhere in the world.

The Act arose because of protests from local farmers and fisher-men in the East Riding of Yorkshire, who had begun to notice the rapid decline of seabirds on nearby Flamborough Head and Bempton Cliffs. A contemporary article in the *Manchester Guardian* revealed that over 100,000 seabirds had been killed in just four months – the adults shot as they sat on their nests, with the chicks subsequently starving to death. The fishermens' motives for saving the birds were not entirely altruistic. The local MP, Christopher Sykes, noted that not only did the seabirds help fishermen find shoals of fish by gathering above them, but also, in foggy weather, the constant cacophony of cries from the cliffs would help guide the fishing boats safely ashore.[38]

The pioneering 1869 law was followed, during the next few decades, by several other important pieces of legislation.[39] But although these did reduce the killing of birds in Britain, they were powerless to prevent the continued importation of skins and feathers from abroad. To stop this, something more was needed. So, in what today's campaigners would recognise as an early use of 'nudge theory', social pressures were brought to bear on the women whose desire to show off the latest fashions was leading to the worldwide massacre of birds.

This began, as befits Victorian high society, with three great British traditions: drinking tea, going to church, and writing letters. During 1889, two separate groups of women began to gather together in genteel drawing-rooms in Surrey and Manchester, to discuss how to solve the problem of the plume trade. One, in Croydon, was called the Fur, Fin and Feather Folk; the other, in the Manchester suburb of Didsbury, adopted a more workmanlike name: the Society for the Protection of Birds, or SPB.

The Fur, Fin and Feather Folk was co-founded by Margaretta Louisa Smith, who on marrying her husband Frank Lemon three years later adopted the wonderfully bizarre moniker Etta Lemon.[40] Born in 1860, Etta was the daughter of an army captain turned Christian evangelist, and from an early age abhorred any cruelty to animals. By the late 1880s, she had joined forces with an older and more experienced animal rights campaigner, Eliza Phillips, who had already been very active in the Society for the Prevention of Cruelty to Animals (later the RSPCA). The early meetings of the Fur, Fin and Feather Folk took place in Phillips's home.

Two hundred miles to the north-west, in Manchester, another formidable woman, Emily Williamson, was also planning an active campaign against the plume trade, focused particularly on the urgent plight of the great crested grebe. Williamson summed up her philosophy thus: 'Women are mostly timid in inaugurating anything, but they are very ready to give their help to a good cause when they are shown the way.' This contrasts with the more uncompromising approach of Etta Lemon. 'The emancipation of women has not yet freed her from slavery to so-called "fashion",' she wrote militantly, 'and nor has a higher education enabled her to grasp this simple question of ethics and aesthetics.'[41]

The two groups used similar tactics. They would attend their local church services on a Sunday, observe which of their friends and acquaintances were sporting feathers in their hats, and then write polite yet firm letters explaining why this was cruel, and how birds had suffered and died in order to satisfy these women's vanity.

Whether by causing the wearers embarrassment, alerting them to their presumably unintended cruelty, or a combination of both, the message soon began to get through. Having been converted to the cause, the women were then persuaded to join their local society, with the aim of taking on the plume trade and ultimately bringing it to an end.

Soon afterwards, in 1891, the two groups decided they would be better off fighting under one banner – the Society for the

Protection of Birds. Emily Williamson became Vice-President, while Etta Lemon took on the role of Honorary Secretary. The first president of the new Society was an equally impressive – and even more socially influential – woman: Winifred, Duchess of Portland, who held the post for over sixty years, until her death in 1954.

In 1904, fifteen years after it was founded, the Society was granted official recognition by King Edward VII – whose mother Queen Victoria and wife Queen Alexandra were early supporters of the cause – and became the *Royal* Society for the Protection of Birds, or, as it is now widely known, the RSPB.

On the other side of the Atlantic, another determined group of women was also campaigning against the savageries of the plume trade. As in Britain, the movement was also centred on high-society drawing rooms, and, once again, afternoon tea parties were the catalyst for the crusade.

In early 1896, in the city of Boston (the setting for another infamous, tea-related protest more than a century earlier) two high-society women, Harriet Lawrence Hemenway and her cousin, Minna B. Hall, organised a series of social events. At these exclusive soirées, while sipping tea and nibbling at sandwiches and cakes, they informed their friends and acquaintances of the terrible slaughter of birds for the fashion trade. Their campaign proved to be highly successful: more than 900 women signed up to the cause, agreeing not just to boycott the wearing of feathers in their hats, but to persuade their friends and acquaintances to do likewise.[42]

Encouraged by this wave of support, Hemenway and Hall founded the Massachusetts Audubon Society later the same year.[43]

Little by little, thanks to their determined stand, the climate of opinion began to shift in favour of the abolitionists, with their campaign also gaining traction in the popular press. In October 1897, the *Chicago Daily Tribune* called on women to save the birds by making a pledge that they 'would not wear birds or bird plumage of any kind except ostrich plumes on their hats'.[44]

But until federal laws were passed to prevent the mass killing of birds at home and abroad, this cruel but very lucrative trade was likely to continue. The next major step came in 1900, when the Lacey Act went into law, prohibiting the transportation of bird skins and feathers across state boundaries. This closed a loophole which until then had allowed hunters to circumvent the bird protection laws in each state simply by taking their ill-gotten gains across the nearest border and selling them there.[45]

The next two decades saw the passing of two more significant laws. The first, the Weeks–McLean Act of 1913, banned the shooting of migrating birds. It was supported by the pioneering car maker Henry Ford, who later wrote that 'The only time I ever used the Ford organisation to influence legislation was on behalf of the birds, and I think the end justified the means.'[46] The Weeks–McLean Act was followed, five years later, by the Migratory Bird Treaty Act of 1918, which protected all native migratory species of North America – and indeed, still does, despite several attempts by former president Donald J. Trump to sidestep it.[47] This final, all-encompassing piece of legislation closed most of the loopholes, by making it 'unlawful to pursue, hunt, take, capture, kill, possess, sell, purchase, barter, import, export, or transport any migratory bird'.[48]

Back in Britain, a combination of campaigning and legislation was also gradually getting results, though not without the occasional setback. In July 1920, the Plumage Bill was put to the vote in the House of Commons but, despite several famous and influential signatories including authors Thomas Hardy and H. G. Wells, failed to pass.[49] By the time it did reach the statute book the following year, the demand for ornate feathered hats was already on the wane – thanks, in a rather delicious irony, to a change in fashion.

In 1915, the American ballroom dancer Irene Castle, who, along with her husband Vernon, had been highly influential in the creation of modern dancing, needed to go into hospital to have her appendix removed. Before the operation, she decided to cut her long hair short, to make it easier to wash while she was in recovery. Afterwards, when she reappeared at social engagements wearing a turban, Castle was persuaded to remove it to reveal her new hairstyle. Her adoring fans, as they say, went wild, and the 'bob' was born.[50]

By the early 1920s, virtually all trendsetting actresses and dancers – the 'social influencers' of their day – wore the bob, which simply did not work with large, feathered hats. The plume trade, which had made such vast profits from adorning women's headgear with wild birds' feathers in what has come to be dubbed 'the Age of Extermination',[51] was finally consigned to a bygone age.[52]

One of the most fascinating aspects of this whole story is the prominent part played, on both sides of the Atlantic, by women. But this is not a simple tale of female triumph against the odds, as it is sometimes portrayed.

Women were not just the founders of the organisations protesting against the trade: they also did the bulk of the campaigning legwork. Yet the prominent men they enlisted to their cause, such as William Brewster in the US, and Alfred Newton and W. H. Hudson in the UK – were usually the public face of their campaigns, often being given the credit for any successes. Moreover, much of the anger and criticism from campaigners and journalists was aimed not at the men who actually *killed* the birds, or at those men who profited from this grisly trade, but only at the women who dared to wear the finished product.[53] In 'Feathered Women', a leaflet published by the newly formed Society for the Protection of Birds in 1893, Hudson himself condemned women who wore feathers as the 'bird-enemy', and even went as far as to suggest that they would fail to attract a husband.[54]

Likewise, a *New York Times* article of July 1898, entitled 'Murderous Millinery', pinned the blame squarely on the 'tender-hearted woman' for whose gratification 'millions of birds have been slaughtered'. The writer went on to condemn 'feather-headed women', who 'invite such public stigma by exhibiting themselves as they do in the relics of murdered innocence'.[55] In this climate of misogyny, the plume hunters – all of whom were men – appeared to be escaping most of the criticism.

Following the defeat of the 1920 Bill, even the author and feminist Virginia Woolf wrote an essay which began by criticising the women who continued to wear bird plumes,[56] to the point of savaging one window-shopper for having 'a stupid face . . . [with] something of the greedy petulance of a pug dog's face at tea time'. 'When she comes to the display of egret plumes, artfully arranged and centrally placed,' wrote Woolf in righteous irony, 'she pauses . . . For, after all, what can be more ethereally and fantastically lovely? The plumes seem to be the natural adornment of spirited and fastidious life, the very symbol of pride and distinction.'

Finally, however, Woolf did direct her anger at the men who drove the whole business – the plume hunters and traders – along with the exclusively male Parliament that had initially rejected the bill, to highlight the double standards in society which condemned women for satisfying their wants and needs through fashion, while praising – or at least tolerating – men's urge to hunt, kill and make a profit.[57]

Meanwhile, another group was suffering terribly from the whims and desires of their richer peers. These were the poor, uneducated, working-class women whose job it was to pluck the feathers from the skins and prepare them for use in the millinery trade. Until Tessa Boase delved into the story in her 2018 book, *Mrs Pankhurst's Purple Feather*,* the plight of these unfortunate women had not been given the coverage it deserves.

Boase tells the sad story of Alice Battershall, a twenty-three-year-old employee at a 'feather factory' in the City of London. In September 1885, Alice was accused of stealing two ostrich feathers, a crime for which she was sentenced to six weeks' hard labour: a fate that stands for the desperation of thousands of young women who worked long hours, and for ridiculously low wages, in what was one of the most profitable industries in London. The feathers she stole, and which her mother sold on for just a shilling each,

---

* Now republished in paperback as *Etta Lemon: The Woman Who Saved the Birds.*

were, when displayed on a ladies' hat, worth as much as £5 (close to £700 at today's values).[58]

Perhaps the most unusual aspect of the whole story, however, is how two groups of women who might be expected to have joined forces to overcome the vested political and economic interests of men – the campaigners against the plume trade, and the suffragettes, campaigning for women's right to vote – turn out to have been implacably opposed to one another.

The title of Boase's book refers to the early suffragettes' tendency to wear the latest and most feminine fashions – including, of course, the feathers of wild birds – to counter the oft-quoted charge that those who desired woman's suffrage were in some ways not 'real women'. This led to a political and social gulf between the two groups of campaigners. On one side there were those like Emmeline Pankhurst, who fought for women to get the vote; on the other, women like Etta Lemon, who wanted to bring an end to the use of feathers in women's fashion. The confrontation was given added spice by the fact that Etta Lemon was herself an 'anti-suffragist', adamantly opposed to the very idea of women being given the vote.[59]

Despite their obvious differences, both groups did ultimately achieve their aims – by using very similar tactics, which could be summed up as 'protests, pamphlets and persuasion'. Indeed, the environmental historian Dr Rob Lambert conflates the two movements in his memorable description of the protestors against the plume trade as 'ornithological suffragettes'.[60]

And just as the women's suffrage movement changed politics for ever, so the achievements of those early campaigners against the plume trade were both crucial and long-lasting. Their triumph would herald the rise of one of the most important movements in global history: the drive to protect wild creatures, and the special places where they live. As Brigid McCormack, Vice President of the National Audubon Society, has observed, this was the first time a popular movement came together to defend birds and the wider environment; something we take for granted today.[61]

During the early decades of the twentieth century, it wasn't just the birds that were going from strength to strength; the cause of bird protection was too. In 1905 the various state and local Audubon Societies came together under the umbrella of the National Audubon Society. Today, known simply as 'Audubon', the society comprises almost 500 independent local chapters, with over 600,000 members. In the UK, the RSPB has grown into one of the largest bird conservation organisations in the world, with over 1.1 million members.[62] Its remit has also broadened from the original concern for the welfare of birds under threat from the plume trade to encompass much broader global issues, including habitat conservation in the UK and abroad, energy and transport use, the loss of biodiversity, and of course the climate crisis.[63] Etta Lemon – who continued to serve and support the RSPB for well over half a century, before her death in 1953 – and Emily Williamson, the forgotten heroine in this story, would surely have been both amazed and proud of what the organisation they helped to found has become.

*

Bird protection has come a very long way since that day in July 1905 when a brave man was killed in cold blood simply for trying to protect a breeding colony of snowy egrets.

Guy Bradley has certainly not been forgotten, though: in 1930, the author and conservationist Marjory Stoneman Douglas published a short story about him,[64] while a 1958 film, *Wind Across the Everglades*, starring Burl Ives and Christopher Plummer, was based partly on his life and untimely death.[65] There are also a number of awards named after him.[66] In a 2013 documentary film, *Guy Bradley, America's First Environmental Martyr*, the presenter Stuart McIver's verdict was that Bradley's martyrdom was the turning point of the fight against the plume hunters. 'It seems funny to have to say this, but [Bradley] probably did more for the cause by giving up his life, than if he had continued as before, if he just kept on being a warden for another fifteen or twenty years'. That may well be true, though it would have been scant comfort to Bradley's grieving widow and children. It also inadvertently implies that we have seen an end to the killing of conservationists trying to defend the natural world and its diminishing resources. Sadly, nothing could be further from the truth.

In September 2021, a BBC News report revealed that in the year 2020 alone, 227 environmental activists around the world – a record number – had been deliberately targeted and killed.[67] The victims included the sixty-five-year-old South African Fikile Ntshangase, shot dead in her home by unknown killers for campaigning against the extension of an opencast coal mine. Also murdered was Óscar Eyraud Adams, gunned down outside his home in Baja California, Mexico, in September that year, while campaigning for indigenous peoples' water rights.

Colombia, which boasts more different species of bird within its borders than any other country on Earth, also holds a less enviable record. More conservationists and environmentalists are killed there than anywhere in the world: sixty-five in 2020 alone – more than a quarter of the global total.[68] One young Colombian, Francisco Vera, has received a number of death threats because of his campaign to save his nation's unique and threatened wildlife. As his terrified mother, Ana Maria Manzanares, says, 'It hurts him. The tranquillity and life we had before are not coming back.' Francisco is just twelve years old.[69]

And, at the heart of this story of global bird protection and conservation, that small, dazzlingly white, waterbird – how has the snowy egret fared in the century since the plume trade finally came to an end?

As early as the 1920s, the breeding colonies of egrets and other waterbirds that had been so devastated by wholesale massacre were beginning to recover. In his 1924 book *Tales of Southern Rivers*, the US adventure writer Zane Grey enthuses about a visit to a colony near Cape Sable, on the western edge of the Everglades:

> Though we saw birds everywhere, in the air and on the foliage, we were not in the least prepared for what a bend in the stream disclosed. Banks of foliage as white with curlew [*sic*] as if with heavy snow! With tremendous flapping of wings that merged into a roar, thousands of curlew took wing, out over the water . . . It was a most wonderful experience.[*][70]

---

[*] We must presume that Grey was a better novelist than he was a naturalist, and that the white birds he saw were not curlews, but snowy egrets.

Today, the snowy egret can be found in wetland habitats across a broad swathe of the southern and central United States. After reaching a peak between 1930 and 1950, following the ending of the plume trade, numbers have fallen in recent years, owing to habitat loss, water pollution, droughts, human disturbance and a reduction in the availability of food. However, even this diminished population represents a conservation success story, especially compared to other once-abundant North American birds such as the passenger pigeon, Carolina parakeet and Eskimo curlew, all of which were ruthlessly hunted to total extinction. By managing to avoid their fate, and giving birth to conservation movements throughout the world, the snowy egret offers a counterweight to another bird in this book, the dodo. The egret's continuing presence in the wetlands of the Americas is a sign that, given enough will and hard work, we can perhaps triumph over human greed, in favour of nature.

8

BALD EAGLE

Haliaeetus leucocephalus

The world is grown so bad, that wrens make prey where eagles dare not perch.

William Shakespeare, *Richard III*, Act I, Scene III

Arms stretched wide, face contorted with hatred and fury, this man means business. In front of him, blocking his path into Washington's Capitol building, stands a police officer, one of a handful trying, in vain, to hold back the baying mob. Almost all the protestors are carrying or wearing the symbols of their cause: the orange beanie hats of the hard-right Proud Boys movement, several Nazi swastikas and, waving in the chill winter wind, the now discredited red, white and blue flag of the Confederacy.[1]

But of all the symbols on display, one of the most curious is emblazoned on the T-shirt worn by that angry protestor. On the front is a large letter Q (representing the bizarre QAnon conspiracy theory), entirely surrounding the image at the centre of the design. Staring back into the camera, with piercing yellow eyes, is the United States' unofficial national bird: the bald eagle.[*]

The unprecedented storming of the US seat of government in January 2021 was not the only occasion, in these turbulent times, when the bald eagle has been invoked by the far-right as a symbol

---

[*] As the Pulitzer Prize-winning author Jack E. Davis points out, the bald eagle has never actually been officially adopted as the USA's national bird. Yet all evidence suggests that this is how it is regarded by the vast majority of Americans, and indeed around the world. See *The Bald Eagle: The Improbable Journey of America's Bird* (New York: Liveright Publishing/W. W. Norton, 2022).

of national supremacy. At the height of the bitterly fought 2020 US election, the campaign to re-elect President Donald Trump launched an official T-shirt featuring an eagle with wings outspread, perched above a circular US flag, topped by a seemingly uncontroversial phrase: 'America First'. Yet this combination of the eagle and that slogan was dog-whistle politics at its most cynical.

The T-shirt design was immediately condemned for its striking similarity to the symbolism used by another powerful right-wing political grouping. As the progressive Jewish Twitter account Bend the Arc pointed out: 'Trump & Pence are proudly displaying a Nazi-inspired shirt on their official campaign website. They are promoting genocidal imagery yet again – just days after President Trump retweeted a video of a supporter chanting "white power".'[2] It went on to explain the tarnished history of the slogan on the T-shirt, which had come to prominence some eighty years earlier in the name of the America First Committee, launched in 1940 to argue against – and ultimately try to prevent – US involvement in the Second World War. Fronted by a bona fide American hero, the pioneering aviator Charles Lindbergh, the movement attracted more than 800,000 paying members, and at one time looked as if it might actually succeed in stopping the USA from entering the war. Then, on 7 December 1941, America First's campaign came to an abrupt end when Japanese forces bombed Pearl Harbor.[3]

The original America First movement was undoubtedly fuelled by anti-Semitism and fascism. In many people's view, therefore, when Donald Trump resurrected the slogan in his inaugural address as US President in January 2017, he knew exactly what he was doing:

linking his own political philosophy to what has been described as the 'hateful legacy' of previous attempts at isolationism.[4]

Not surprisingly, Trump's campaign spokesman was quick to deny any similarity between the image on the America First T-shirt and the symbolism of the Nazis, while simultaneously mocking his opponents and making an overt appeal to old-fashioned American patriotism: 'This is moronic. In Democrats' America, Mount Rushmore glorifies white supremacy and the bald eagle with an American flag is a Nazi symbol. They have lost their minds.'[5]

Such fervent denials lose all credibility when we take a closer look at the image of the eagle on that notorious T-shirt and, more specifically, the direction it is facing. The Great Seal of the United States shows the eagle facing left (as the viewer looks at it), as do the vast majority of historical and present-day examples of eagle iconography, including the current German coat of arms. Yet the Trump campaign chose to depict the eagle facing *right* – the same direction as the infamous Nazi symbol, and the image still used today by various Neo-Nazi groups.[6]

So, was the way the eagle on the T-shirt is depicted – facing right, rather than left – purely a coincidence, or does it hint at something more sinister? The US historian Steven Heller, who specialises in decoding right-wing iconography and symbolism,[7] has no doubt it reveals a threatening intent, cleverly hidden behind an apparently innocent appeal to patriotism:

I find it hard to believe that the direction of the eagle's head (mandated by Harry Truman) was [*sic*] flaunted by Trump's designers without knowing the symbolism. But his gang know what they're doing. They understand the force of well-staged performance, props and all. So, my belief is that this eagle is indeed a nod (a secret handshake, so to speak) between Trump and racist America.[8]

Donald Trump is not the only leader of a powerful nation or regime to have followed the Nazis' lead in putting the eagle at the head of his movement's iconography. When General Franco seized power in Spain after the bitter civil war, he placed the ancient heraldic symbol of the Eagle of St. John[9] – previously associated with the fifteenth-century 'Catholic Monarchs' Ferdinand and Isabella – on the nation's flag, while another dictator, Saddam Hussein, chose the Eagle of Saladin, a widely used symbol of Arab nationalism. And in 1993, following the sudden and rapid break-up of the Soviet Union, the Russian Federation resurrected an eagle symbol dating back to the late sixteenth century. Its coat of arms depicts a golden two-headed eagle on a red background.[10]

The eagle is the national bird of more countries and nation states than any other: these include Albania, Germany, Indonesia, Kazakhstan, Mexico, Namibia, Panama, the Philippines, Poland, Scotland, Serbia, Zambia, Zimbabwe and, in reality if not officially, the United States.[11] It is also depicted on the flags of Albania, American Samoa, Egypt, Kazakhstan, Mexico, Montenegro, Moldova, Serbia and the US Virgin Islands, as well as those of a number of US states.[12]

As Professor Janine Rogers, of Canada's Mount Allison University, points out, the cultures where these symbols originated, and most modern uses of the bird, such as those on flags, are not generally linked with totalitarian regimes. However, as she goes on to warn, the eagle 'carries with it an inherent sense of menace that seems to adapt all too easily to tyranny and oppression'.[13] Indeed, its use has always been problematic, as the eagle's long and eventful history, both as a symbol, and as a real, living bird, reveals.

It could be argued that, in reality, there is no such a thing as an eagle.* The name 'eagle' is randomly attached to several sub-groups of large raptors, including hawk-eagles, snake-eagles, buzzard-eagles, serpent-eagles and fish-eagles (including the bald eagle), as well as the 'true' eagles of the genus *Aquila*. So, in truth, the name 'eagle' is not so much a biological fact, more of a linguistic convenience.

What these birds do have in common is that they are mostly very large and obvious, and at or near the top of their food chain. Eagles can be found in all six inhabited continents, ranging from far beyond the Arctic Circle, through the temperate and tropical zones, to the equator and beyond. They have also managed to adapt to a very wide range of habitats, from the highest mountains, via coastal and freshwater wetlands, woods, forests and grasslands, to the hottest deserts.

The bald eagle itself is equally catholic in its choices of where to live. It can be found across virtually the whole of North America: from the riverine forests of Alaska and Canada in the north, throughout the whole of the lower forty-eight states, to northern Mexico in the south. In Arizona, one population even nests in the

---

* Granted, all the various species which have eagle as their name are in the same family (*Accipitridae*, which also includes buzzards, hawks, harriers, kites, and Old World vultures), but they are not otherwise very closely related to one another.

hot and arid desert. The highest densities of bald eagles are found in the wilderness of Alaska, but I have seen a pair – perched on their enormous nest – on a suburban housing estate in Fort Myers, Florida. In a rather surreal scene, they went about their day-to-day lives apparently oblivious to the comings and goings of the local residents, who mostly ignored them, too.

The first thing to note about the bald eagle is that, unlike the Old and New World vultures, which do lack feathers on their heads, it is not actually bald. The name comes from the pure white colour of the adult bird's head, which from a distance appears as if it is unfeathered. This – and the bright yellow bill and legs – make the bald eagle easy to tell apart from the only other member of its family found in North America, the golden eagle.

The bald eagle – which, apart from the very rare California condor, is the largest bird of prey in North America[*] – is one of ten species in the genus *Haliaeetus*, from the Greek meaning 'sea-eagle'. Its sibling-species is the white-tailed eagle of northern Europe and Asia, while another close relative is the African fish eagle, which also has a bright white head. Like other members of its genus, the bald eagle's main diet is fish, which it catches by swooping down to grab the victim from the surface of the water with those sharp talons, stealing them from other birds like the osprey or peregrine (a behaviour known as kleptoparasitism), or scavenging them, for instance when salmon wash up exhausted

---

[*] Like most raptors, the female is noticeably larger than the male: the body-length varies from 70–102 cm, the wingspan is between 1.8 and 2.3 metres, and an adult eagle weighs between 3 and 6.3kg. See J. Ferguson-Lees and D. Christie, *Raptors of the World* (London: Christopher Helm, 2001).

on the banks of rivers after their journey from ocean to spawning ground.* Like many birds of prey, bald eagles are also opportunistic hunters: when the chance presents itself, they will catch a wide range of mammals, reptiles and birds, including grebes, coots, gulls and loons (divers). They have been known to feed on more than 400 species in all, including a beached whale; in terms of a varied diet, among America's raptors they are second only to the red-tailed hawk.[14]

Where the bald eagle triumphs over all its rivals is in the size of its nest. This huge structure is the largest arboreal nest of any bird in the world, trumped only by the enormous mounds constructed by the ground-nesting mallee fowl of Australia.† One nest, near St Petersburg in Florida, measured 2.9 metres wide and 6 metres deep, and was thought to weigh at least 2 tonnes, earning it a well-deserved place in the *Guinness Book of World Records*.[15] So there can be no doubt that, like other members of its family, the bald eagle is a very impressive bird, even if its habits – especially the kleptoparasitism and scavenging – leave something to be desired. Yet those habits did not stop the Founding Fathers of the United States of America from adopting the bald eagle as the *de facto* national bird.

*

* Eagles also frequently take the remains of salmon and other fish killed by grizzly bears, wolves and foxes, sometimes following these predators in order to take advantage of this easy supply of food. R. Armstrong, 'The Importance of Fish to Bald Eagles in Southeast Alaska: A Review' (US Forest Service).

† The mallee fowl's nest can measure 4.6 by 10.6 metres and include material weighing up to 300 tonnes.

On 20 June 1782, some six years after independence from Britain had been declared, the Irish-born Secretary to Congress, Charles Thomson, unveiled the final design for the Great Seal. Eventually this would become the national emblem of the United States of America, since when, as the Pulitzer Prize-winning author Jack E. Davis put it, 'no animal in American history . . . has to the same extent been the simultaneous object of reverence and recrimination.'[16] A fierce-looking bald eagle was depicted with outstretched wings, beneath a shield with thirteen stars (representing the original colonies that fought for independence), and holding a scroll with the slogan *E Pluribus Unum* (Latin for 'Out of many, one') in its bill. In its talons, the bird carried an olive branch and a bundle of arrows – a rather ambiguous pairing of peace with war. As Thomson's co-designer William Barton noted, 'The Eagle displayed is the Symbol of Supreme Power and Authority.'[17] The final design, passed by Congress later that day, had emerged after a long and heated debate among the Founding Fathers of the nation, and a convoluted selection process, with three separate committees, had discarded a variety of symbols, including the eye of providence in a triangle, a pyramid, and the year 1776 in Roman numerals.

Incredibly, the eagle did not appear in the design at all until the last minute, when it was suggested by the third and final committee. Even then, it did not yet sport the distinctive snow-white head of the bald eagle. Thomson's persuasive manner and detailed explanation of the eagle's long-standing symbolism, representing qualities such as strength and courage, together with the fact that the bald eagle was definitively American, rather than European, finally won the approval of Congress. It is tempting to speculate that, after so many years of fruitless discussion, they might have accepted almost any design.[18]

Not everybody was happy. Famously, a key dissenting voice came from the elder statesman of American politics, Benjamin Franklin. Having reached the venerable age of 76, Franklin was in no mood to rubber-stamp the new symbol. 'For my own part', he wrote to his daughter Sarah, in January 1784, 'I wish that the bald eagle had not been chosen as the representative of our country. He is a bird of bad moral character. He does not get his living honestly.'[19] Franklin went on to explain his thinking: for him it was precisely the eagle's scavenging and thieving that were the problem: the way it sits in a tree and watches the 'fish-hawk' [osprey], then, when that bird manages to catch a fish, swoops down to steal it. Franklin also accused the eagle of being 'a rank coward: the little king bird not bigger than a sparrow attacks him boldly and drives him out of the district.'[20] In a lasting blow to the eagle's reputation, he went on to suggest that another species, the wild turkey (see chapter 3), might have been a better choice. 'For in truth, the turkey is in comparison a much more respectable bird . . . a bird of courage and would not hesitate to attack a grenadier of the British guards, who should presume to invade his farmyard with a red coat on.'[21]

Later commentators have suggested that this has been blown out of proportion, and that Franklin made no such objections to the eagle at the time it was approved.[22] It is also quite possible that he was writing with his tongue in his cheek.[23] Nevertheless, the seeds of doubt had been sown. They were magnified half a century later when the celebrated bird artist and explorer John James Audubon came out in apparent agreement with Franklin. 'Suffer me, kind reader,' he wrote, 'to say how much I grieve that [the bald eagle] should have been selected as the emblem of my country.'[24]

One myth that has gained widespread currency is that the eagle was originally chosen because, during one of the first battles of the War of Independence, eagles circled the heads of the soldiers and uttered their raucous, high-pitched cries, an act interpreted as 'shrieking for freedom'. As the early-twentieth-century children's author Maude Grant concluded: 'Thus the eagle, full of the boundless spirit of freedom . . . has become the national emblem of a country that offers freedom in word and thought and an opportunity for a full and free expansion into the boundless space of the future.'

This was not the first time, and it would not be the last, that the symbolism of the mighty eagle, soaring free, would be invoked to justify the idea of the United States as the world's primary bastion of freedom and democracy. The reality is rather more complicated, as many people around the world who have suffered grievously under America's influence would attest.

In today's America, the symbol of the bald eagle is so ubiquitous, so commonplace, and in many ways so clichéd, that it is easy to overlook its presence. As well as being on the Great Seal of the President of the United States (used to authenticate official federal government documents), it appears on the reverse side of the one-dollar bill – of which there are almost 12 *billion* in circulation at any one time. Elsewhere, it features on flags, military uniforms, public monuments and buildings, passports, and other official documents issued by the United States government. Throughout the world, it is prominently placed on the gates and doors of all US embassies and consulates. In a less elevated form, it has appeared on the cover of Marvel comic books, including *Superman* and *Captain America*, and as 'a hood ornament, door knocker, money clip, and chest tattoo – the stuff of trinkets and gewgaws'.[25]

Perhaps most famously of all, a bald eagle featured as the 'mission patch' on the spacesuits worn by the three Apollo 11 astronauts in July 1969 during their historic journey to land on the Moon. In a fortuitous link with the dove (see chapter 2), this image featured the bird carrying an olive branch, echoing the words on a commemorative plaque placed on the Moon's surface: 'We came in peace for all mankind.'[26] The Apollo 11 lunar module was also, appropriately, named *Eagle*, giving rise to one of the best-known quotations of this, or any other age: 'The *Eagle* has landed.'*

The Founding Fathers of the United States were not the only political grouping to choose the symbol of the eagle to justify and promote their own power and authority. Indeed, it is quite difficult to find a major historical empire or civilisation that has *not*, at one time or another, used an eagle in this manner. Many of the indigenous peoples of North America regarded the bald eagle as a sacred bird: the 'bird of heaven', a spiritual messenger between gods and human beings, and as the 'Thunderbird', a supernatural being which, as its name suggests, brings thunder, lightning and life-giving rain.[27]

Elsewhere in the Americas, two major (yet ultimately doomed) civilisations – the Aztecs and the Mayans – placed the eagle at the

---

* This line subsequently became even more famous when appropriated by Jack Higgins for the title of his 1975 counterfactual thriller about a Nazi plot to assassinate Winston Churchill – perhaps also a nod to the sinister, rightward-facing eagle.

centre of their mythologies. For the Aztecs, it was both a warrior and a symbol of the birth of the sun; hence the bird's blackened wingtips, which were said to have been scorched by the sun's heat.[28] The Mayans chose a more ambiguous, double-headed version, representing the eternal struggle between good and evil. And in Australia, where three species of eagle occur, the largest, the wedge-tailed, is central to Aboriginal cultures, often paired in an antagonistic relationship with the crow.[29]

As you might expect, there are many references to eagles in the Bible. In the Old Testament, Saul and Jonathan are described admiringly as being 'swifter than eagles',[30] while in the final book of the New Testament, Revelation, the fourth apocalyptic beast to appear is 'like a flying eagle'.[31] Eagles were also at the centre of three of the world's most influential early Old World civilisations: the Greek, the Persian and the Roman empires, which spanned the period from the eighth century BC to several centuries after the birth of Christ. In all these ancient societies, eagles were regarded as symbols of supreme power and authority, owing to their obvious qualities of strength, longevity and freedom. Their position as a top predator, and their habit of rising high into the sky (using thermal air currents to gain height, and so appearing more effortless than other flying birds), gave them an obvious affinity with gods and, by later association, with mortal kings and emperors who wished to suggest that they shared those heavenly powers.* The Ancient Greeks, indeed, explicitly chose the eagle as the symbol of Zeus, the

---

* Although ironically, the eagle's position as the 'King of Birds' was subverted by the humble and tiny wren; a tale told in many northern hemisphere cultures. See Stephen Moss, *The Wren: A Biography* (London: Square Peg, 2018).

Father of Gods and Men; the closest equivalent to an all-powerful being in their polytheistic society.[32] An eagle is supposed to have brought nectar to Zeus when he was a baby in a cave; later, he is said to have fought alongside it, and to have sometimes taken the form of an eagle when in battle.[33]

The first Persian (or Achaemenid) Empire began later than its Greek counterpart, and lasted for a much shorter time: just over two centuries, from roughly 550 to 331 BC. It covered an area of 5.5 million square km – roughly twice the size of present-day Argentina – making it the largest empire the world had seen at that point in human history.[34] The empire's founder, Cyrus the Great, carried a flag showing a 'shahbaz', a mythical bird likely to represent either a golden or an eastern imperial eagle, which, like the bald eagle on the US Great Seal, was always depicted with wings outstretched.

Cyrus himself was regarded as a model ruler: brave and powerful, as you would expect, but also tolerant, efficient and magnanimous to those weaker than him. Like Zeus, he balanced strength with more benevolent qualities, and was much admired, not just amongst his own people but far beyond, in the Greek Empire.[35] Until this point, the symbolism of the eagle appears to be carefully balanced, between its power, strength and authority on the one hand, and the more nuanced qualities of wisdom and compassion on the other.

Pinning down exactly when, how and especially *why* this changed to something more troubling – the eagle as a symbol of tyrannical, rather than benevolent, power – is not easy. My contention is that, although it took many centuries to find its full manifestation, this shift in symbolism began just over 2,000 years ago, at the height of the Roman Empire.

In the 1979 film *Monty Python's Life of Brian*, Reg, the leader of the fictional People's Front of Judea (played by John Cleese) famously asks, 'What have the Romans ever done for us?'[36] But perhaps a more relevant question might be, 'What did the Romans ever do for *themselves*?' Few empires, before or since, have been quite so ruthless in their approach: first conquering, and then taking social and economic advantage of, the lands and peoples it had won. And the Romans did so with little or no compunction for the welfare of the people they had subjugated. So, when the Romans used the symbol of the eagle, primarily as an expression of military might and power, did it now represent a darker, and more sinister, meaning than it had with previous civilisations?

Eagles had previously appeared as ensigns, on flags and standards carried into battle by various great powers, but the Romans went one better: they created a motif designed to strike fear and terror into the hearts of their enemies, and act as a spur and incentive to their own forces. Its prominence was the brainchild of the general and statesman, Gaius Marius.

The key change came after a devastating military defeat at the Battle of Arausio, in what is now south-east France, on 6 October 105 BC. The two Roman armies had been expected to triumph easily against far weaker opposition: two northern Germanic tribes, which supposedly lacked the iron discipline and huge material resources of the imperial forces. But a long-standing rivalry between the Roman commanders proved to be their undoing: as a result, an estimated

80,000 of their soldiers, and perhaps half as many administrators and camp followers, were killed.

The repercussions of this shocking defeat allowed Gaius Marius his chance to become the saviour of Rome. One of his first actions was to ditch four of the five animal-based military ensigns – the ox, wolf, horse and boar – retaining only the fifth: the eagle, or *aquila*. Made of silver or bronze and fixed to the top of a long pole so that it could be widely seen by both officers and men, this was carried into battle by a single standard-bearer, or *aquilifer*, in each legion. His only job was to safeguard this precious object; if necessary, by sacrificing his own life.

The switch instigated by Gaius Marius, from five symbols to just one, may appear relatively minor, but it helped stem dissent and rivalry, and united the entire Roman forces, as they showed their loyalty to this one, all-powerful, symbol.[37] It is hard to over-emphasise just how significant this was to the Roman army. Each soldier regarded the *aquila* as a sacred symbol, which must be guarded at all costs. If it should fall into enemy hands, then not just the battle but the legion's honour would be lost.

Perhaps the best demonstration of the importance of the *aquila* occurred in 55 BC, when the Romans, led by Julius Caesar, attempted to invade Britain. The story goes that, seeing the British forces massed at the top of the beach, the Roman troops were understandably reluctant to disembark from their ships, and wade through the shallows in heavy armour to reach the shore. According to later accounts, the legion's standard-bearer then seized the initiative, crying out, 'Leap, fellow soldiers, unless you wish to betray your eagle to the enemy; I for my part, will perform my duty to the republic and to my general.'[38] He promptly jumped into the sea and began to make his way ashore, followed, after a brief hesitation, by the rest of the troops.[*] Thanks to stories like this, the eagle, head turned and wings outstretched, became virtually ubiquitous throughout the Roman Empire – not just *a* symbol, but *the* symbol of supreme imperial powers.

Some nine centuries later, at the beginning of the Holy Roman Empire in AD 800, the first emperor Charlemagne (Charles the Great) was crowned by Pope Leo III. To help cement his fragile power base, one of Charlemagne's first acts was to resurrect the original Roman eagle. The design was a familiar one: an eagle with outstretched wings and its head turned to one side. From the mid-thirteenth century, as the empire was beginning to decline, the eagle was often depicted double-headed, lending it an even more dubious aura.[39]

Successive Holy Roman Emperors, including Frederick I (known as 'Barbarossa' because of his red beard), continued to use the eagle symbol, in what might be described as an early form of 'branding'. And it worked. Although the empire eventually lost much of the territory it once ruled, it was not finally dissolved until 1806, following a catastrophic military defeat by Napoleon of the last emperor, Francis II, at the Battle of Austerlitz. The historian Tom Holland points out that Napoleon also adopted the eagle symbol, directly modelling it on the Roman one.[40]

The Holy Roman Empire – which, as the old joke goes, was neither holy, nor Roman, nor an empire – may have had its

---

[*] Ironically, this attempted invasion failed; the successful conquest did not occur until almost a century later, in AD 43, under the Emperor Claudius.

vicissitudes, but it cemented the place of the eagle symbol at the heart of Europe. This would, within a century or so, have unforeseen, and very sinister, consequences.

From time to time, and especially during periods of political instability and change, the eagle became embroiled in controversy. One such dispute began on 11 November 1919, one year to the day after the First World War had ended with the defeat – and subsequent humiliation – of Germany. Friedrich Ebert, the first president of the newly formed German Republic, declared that the imperial eagle would henceforth be the official emblem of the nation.

It was not the choice of the bird itself that caused offence, but the way it had been designed. The stark, angular depiction of a black bird with a bright red bill, tongue and feet was far too modernist in style to win approval from conservative politicians. It was widely derided in right-wing publications and condemned by a largely unknown young politician as a 'Jewish bankruptcy vulture'. His name: Adolf Hitler.[41]

Four years later, in 1923, the cultural historian Arthur Moeller van den Bruck published a book in which he divided the history of the German nation into three sections, which he termed *reichs*, or realms. The first he considered to be the Holy Roman Empire, running from AD 800 to 1806; the second, known as the German Empire, endured a much briefer period from 1871 to 1918, ending with Germany's ignominious defeat in the First World War.[42] But it was the next period, which would become more famous – or indeed, infamous – than any other, whose name appears as the title of Moeller van den Bruck's book: *Das Dritte Reich* – or, as it is better known today, 'The Third Reich'.[43]

Two years after his book was published, Arthur Moeller van den Bruck took his own life in Berlin. Perhaps fortunately, he did not live to see the consequences of his dangerous notion. He would not live to witness how, during the following decade, it inspired an artist, labourer, soldier and politician, Adolf Hitler, to take the concept of a third German Empire, reimagined as a 'Thousand-year Reich', and try to make it a reality.

The Third Reich may have fallen far short of that vaunting ambition – lasting for just 12 years, from 1933 to 1945 – but arguably it had a greater global impact than any other political movement in history. Its menacing and malevolent emblem combined two ancient German symbols, which together would come to define the horrors of Nazi rule: an eagle clutching a wreath with a swastika at its centre.[44]

Like a kleptomaniacal magpie, Hitler ransacked history for any allusion to the former greatness of the German nation which he might use to encourage the rise of German nationalism. So when he and the Nazi Party chose the eagle as their symbol, they knew exactly what they were doing. But, in a crucial change from President Ebert's eagle, the Nazi eagle faced not to the left, as we look at it, but – just like Donald Trump's bird – to the right. The idea was that the bird appeared to be facing east, towards Russia, by then the target of Hitler's military plans.

Soon, the right-facing eagle – now rebranded as the *Parteiadler*, or 'party eagle' – had become ubiquitous. As it had for the

Romans, the eagle helped give the Nazi movement and its adherents a common purpose – an almost mystical higher calling. The historian Justin Hayes describes it as 'the ancient use of a symbol as sympathetic magic', bringing triumph and victory to those who displayed it. 'Perhaps it elevates the wearer to the status of "eagle-bearer", he ponders: 'a class of men who were portrayed as the bravest and most virtuous of legionaries'.[45]

The echo of Roman symbolism was quite deliberate. In 1878, half a century before the Nazi Party's rapid rise to power, the composer and German nationalist Richard Wagner had written an essay entitled '*Was ist Deutsch?*', in which he overtly linked his nation's self-image to that of the Roman Empire: 'In their longing for "German grandeur", Germans can . . . not yet dream of anything other than something similar to the restoration of the Roman Empire. In this *even the most good-natured German* [my italics] is seized by an unmistakable lust for domination and a craving for supreme power over other peoples.'[46]

It is no accident that Hitler put Wagner's philosophy at the centre of his own distorted view of what it meant to be German, with predictably disastrous results. He too was obsessed with the power of the Roman Empire, referring to Rome as 'the best teacher', and 'the greatest . . . political creation . . . of all', for the way in which it 'succeeded in completely dominating all neighbouring peoples'. He also praised the Romans' rigorous military discipline, adapted their imperial salute into the infamous raised-arm Nazi greeting, and promoted their influence in art and architecture. Most importantly of all, he placed the eagle at the centre of Nazi symbolism.[47]

Today, we are still faced with the thorny issue of whether the eagle is a positive or negative symbol. Does it represent the virtues of power used wisely, as in many ancient cultures, or has it, through its deep-rooted association with Nazi Germany and Fascism, inevitably become what Hayes calls a symbol of 'totalitarian governmental power and ruthless ethnocentrism'?[48]

Another view would be that the image of the eagle has long since lost its original connection with the bird itself and become simply a convenient symbolic shorthand for strength and power, adopted without thinking by brand designers. Yet even now, the eagle still has the power to shock, and sometimes backfire on those who choose to use it without considering its controversial past.

In 2007, during Barclays Bank's attempted takeover of the Dutch bank ABN AMRO, it was widely reported that the company had removed a giant metal eagle from the roof of its Dorset branch, apparently in response to concerns from the Dutch about the perceived similarity to the Nazis' eagle symbol.[49] It was suggested that, should the takeover succeed, the eagle would be permanently dropped as the bank's logo.[50] In the end, the bid fell through – and the Barclays eagle remains.

Meanwhile, according to Steven Heller, Nazi iconography, which he refers to as 'the most effective identity system in history',[51] continues to be appropriated by popular culture, notably by designers, rock bands and company logos. He points to what he calls the 'historical amnesia and flagrant misappropriation' of the eagle image by the UK-based clothing brand Boy London, whose 'Heritage Collection' displays a prominent representation of an eagle – facing right – across the front and back of its caps, jackets and hoodies.[52]

This is despite the fact that in 2014, the *Daily Mail* revealed news of a customer boycott of the brand, because of the logo's clear visual links with Nazi Germany.[53] As Heller points out, the company, which has a largely gay male clientele, also uses the slogan, 'The strength of the country lies in its youth' – a phrase eerily reminiscent of the language used by the Hitler Youth movement.[54]

Eagles are also cynically used to appeal to an even younger market. In the United States, the National Rifle Association (NRA) uses an endearing cartoon character, Eddie Eagle, to teach pre-school children about gun safety. All very laudable, until we realise that in 2020 more than 4,300 children and teenagers were killed by guns in the US, making it the leading cause of death for the nation's youth.[55] Studies have shown that, despite the NRA's claims that Eddie helps reduce gun violence towards children, he has had absolutely no measurable effect.[56] It also, as has been widely pointed out, places the onus on ending America's epidemic of gun violence on child victims, rather than adult perpetrators.[57]

It is tempting to assume that these more recent examples of the eagle in iconography, marketing and politics are essentially accidental, lazy or careless references to the Nazi brand, rather than deliberately provocative. And, of course, there are many examples down the ages of the eagle serving a more benevolent purpose, as a sign of wisdom and power used responsibly. Yet to focus solely on these would be to brush under the carpet the other, more troubling, ways in which the eagle has been – and continues to be – invoked. Certainly, the way in which those ultra-right white supremacists storming the Capitol displayed the eagle with such pride, as they called for a 'Fourth Reich' to finish the job Hitler started, should give us pause for thought.[58]

How have the American people, and their state and federal governments, meanwhile, looked after the actual bald eagle since it was first selected as the symbol of the fledgling United States of America? The simple answer would be, 'Not very well at all.'

Like many birds of prey, the bald eagle's relationship with the people who live alongside it has for long been an often troubled one. Like the raven (see chapter 1), eagles essentially compete with humans for natural resources: in the bald eagle's case, fish. As a result, for two centuries or more after Europeans first settled in North America, the bird was ruthlessly persecuted.

It was also one of the first wild birds to send a warning to its human persecutors. As long ago as the early nineteenth century, Audubon wrote of a seriously depleted future for the bald eagle and, by implication, the rest of the natural world: 'A century hence they will not be here as I see them, Nature will have been robbed of many brilliant charms.'[59]

At that time, it has been estimated that there were between 300,000 and 500,000 bald eagles in North America. But in less than forty years, from 1917 to 1953, more than 100,000 eagles were shot in the state of Alaska alone. Others died from lead poisoning, from eating the corpses of other shot animals, or were caught in traps set either to catch beavers and muskrats for their fur, or to control predators such as wolves and coyotes. By 1930, conservationists were warning that the bald eagle was in very real danger of going extinct. One magazine article linked the real with the symbolic bird, with one eminent ornithologist stating that 'In a few years the American bald eagle will be seen only on coins and coat of arms of the United States unless drastic action is taken to save these birds from extinction.'[60]

Even worse was to come. Invented in the nineteenth century, but only discovered to be a powerful insecticide in 1939, DDT was quickly hailed as a 'wonder pesticide', which would eradicate insect pests, and so allow farmers to achieve higher crop yields. But no-one had thought about what might happen as a consequence of its widespread and uncritical use from the mid-1940s onwards.

DDT worked: it killed insect pests very effectively. But it also lingered in the food chain, becoming more and more concentrated the higher up it went. By the time it reached the predators at the very top of the chain – including the bald eagle – it caused huge damage, by thinning the birds' eggshells so much that the chicks failed to hatch.

The fall in the bald eagle population – and in those of other raptor species such as the peregrine – was both rapid and wholly unprecedented. By the early 1960s, barely a decade after DDT use became the norm for US farmers, there were just 412 pairs of bald eagles remaining in the lower forty-eight states. Worse still, almost all were adults, sporting the classic white headdress, which indicated that the chicks were simply not surviving to adulthood. Eagles can live for up to twenty years, so paradoxically their longevity was temporarily masking their almost total lack of breeding success.

Bald eagles were already protected – in theory at least – by the Migratory Bird Treaty Act of 1918,[61] and more specifically by the

Bald and Golden Eagle Protection Act, passed in 1940.[62] In reality, however, these made little or no difference to their persecution and poisoning, which continued apace. Only after the outcry following Rachel Carson's 1962 book *Silent Spring*,[63] which kick-started the modern environmental movement, did the tide finally begin to turn. As Jack E. Davis remarks, 'Twice, not once, the United States nearly lost its flagship bird from the wild, and twice people aided its return.'[64]

Despite having come to the very edge of extinction, the bald eagle bounced back very rapidly indeed. By the early 1980s, just a decade after DDT was finally banned in the US, numbers had risen to an estimated 100,000 birds; ten years later there were about 115,000, the vast majority of which were in Alaska and the Canadian province of British Columbia. As a result, in 1995 the species was transferred from the federal government's list of endangered species to the threatened species list; twelve years later, in July 2007, it was removed altogether.[65]

To celebrate the eagle's remarkable recovery from the brink of extinction, on 20 June each year (the anniversary of the original adoption of the bald eagle as the symbol of the United States), the American Eagle Foundation celebrates 'American Eagle Day'.[66] As the AEF's founder and president Al Cincere says, 'We once almost lost this precious national treasure due to our own mistakes and neglect . . . but we the people joined together, rose to the occasion, and vigilantly brought it back to America's lands, waterways and skies.'[67] Jack E. Davis agrees, in the conclusion of his masterful account of the bald eagle's journey:

In the twenty-first century, *Haliaeetus leucocephalus* achieved the vastness across the continent that it had known before it became America's bird, and a circle closed. In the centuries in between, the species witnessed danger breaking out, the subduing of its eagledoms, and the predatory ways of the people who accused the raptor of heinous crimes. Then, in the presence of its only true predator, the bird of America came to know succour and rescue, a sovereignty restored, and freedom renewed.[68]

He may have spoken too soon. In December 2016, the US Fish and Wildlife Service proposed issuing permits to allow the wind power industry to avoid prosecution over the accidental deaths (from colliding with turbines) of over 4,000 eagles every year for the next thirty years.[69] In December 2019, as one of the last acts of his chaotic presidency, Donald Trump announced major changes to the Endangered Species Act, which had once helped bring the bald eagle back from the brink. These would seriously weaken the Act's powers by allowing economic factors to be considered when deciding whether to permit development that might harm wild creatures and their habitats.[70] And as recently as March 2021, it was reported that the mysterious deaths of several hundred bald eagles in the south-eastern states of the US, from the mid-1990s onwards, had been caused by bromide poisoning, the source of which still remains unexplained.[71]

We know that one of the reasons eagles were originally chosen as symbols of power, by such a wide range of peoples and civilisations, was their wild nature, and their unwillingness to be subservient to us humans. So it is perhaps fitting that in 2015, when the then plain Mr Donald Trump decided to pose with a bald eagle for a *Time* magazine cover story, things didn't quite go to plan. At first, the eagle (aptly named 'Uncle Sam') played along. But as the photoshoot went on, he became more and more agitated, flapping his wings and finally attacking Trump. Poetic justice, perhaps.[72]

When threatened, nature does have a habit of biting back – both literally, as in the case of the eagle and Donald Trump, and metaphorically. And the more omnipotent the human agent, the greater the revenge wrought by wild creatures, as we shall discover in the next chapter.

9

TREE SPARROW

Passer montanus

There's a special providence in the fall of a sparrow.
William Shakespeare, *Hamlet*, Act V, Scene ii

SATURDAY, 13 DECEMBER 1958; SHANGHAI, CHINA

As the new day dawned, huge crowds began to gather in the streets. They teemed through the city, waving thousands of red flags – the symbol of the Chinese Communist Revolution – filling the air with their blood-curdling war cries. The crescendo of noise rose to a deafening pitch, as schoolchildren, students, farmers, factory workers and members of the People's Liberation Army all mobilised against a common enemy.

Soon after sunrise, the slaughter began. While the oldest and youngest kept watch, the others embarked on a mass killing spree, in what one newspaper called 'total war'.

This ragtag army pursued their targets with a single-minded ruthlessness, using poles, nets, traps and guns. Others banged pots and pans in a relentless rhythm, to disturb and confuse their quarry. And all the time they shouted, screamed, cheered and whooped in triumph and delight.

At first, their opponents tried to gather together, seeking safety in numbers. But there was simply nowhere to hide. Gradually, one by one, they fell to the ground and were shot, strangled or simply died from sheer exhaustion.

All over China, these helpless victims died in city streets and rural fields, in public parks and private gardens, on rooftops and in gutters.

Some even fell straight from the sky, before being summarily despatched. By nightfall, in Shanghai alone, almost 200,000 were dead.[1]

We are familiar with horrific accounts of violent genocide. But in this case, the victims of the massacre were not human beings, but sparrows.* Or, as the ruling cadre of the People's Republic of China, led by the all-powerful party chairman Mao Zedong, had branded them, one of the 'Four Pests'.

The idea behind this campaign, part of the wider political and social crusade known as the Great Leap Forward, unveiled in January 1958, was to eradicate four different groups of animals, all denounced as 'vermin'. Colourful posters, including one particularly lurid image of a knife skewering the quartet of intended victims, exhorted China's loyal citizens to 'Exterminate the four pests!' These were rodents, which carried bubonic plague; mosquitoes, which spread various diseases, including malaria; flies, which were both ubiquitous and infuriating; and finally, and most importantly, the sparrows, which ate valuable seeds and grains, and so threatened the annual harvest.

Of the four, sparrows were singled out as the main target in what soon became known, with the Chinese rulers' perennial fondness

---

* Specifically, tree sparrows, members of the family *Passeridae* (Old World Sparrows), though North American readers may be puzzled by the identity of the species concerned. This is the Eurasian tree sparrow *Passer montanus*, not the American tree sparrow *Spizelloides arborea*, which belongs to a different family, the New World sparrows (*Passerellidae*). In Europe, the tree sparrow is a mainly rural bird, while its cousin the house sparrow *Passer domesticus* is found in urban areas. But in China, where house sparrows are far less common, the tree sparrow has adapted to live in both urban and rural habitats.

for slogans, as the Great Sparrow Campaign. Government scientists had calculated that a single sparrow could consume 4.5 kilos of grain per year; therefore, they inferred, for every million sparrows killed, enough would be spared to feed 60,000 people. The maths was theoretically correct; yet the outcome would be the exact opposite of that desired.

After the terrible privations of the previous decade, since Mao had first come to power in 1949, the Chinese people needed all the food they could get. The campaign was guaranteed to be popular in both the cities and countryside, and so would help to unify the nation behind their supreme leader. It saw hundreds of millions of sparrows – along with vast numbers of the other three 'pests', mosquitoes, flies and rodents – successfully hunted down and killed.[*] Nests were destroyed, with eggs and chicks laying smashed and beaten on the ground beneath. As one eyewitness wrote:

> Even those birds that had managed to survive the initial cull were then pursued by villagers and townspeople, banging pots and pans from dawn to dusk, preventing the birds from breeding or roosting, and eventually driving them to perish from sheer exhaustion. A sparrow could be killed in several ways, and all were to be used in this struggle to the death.[2]

Everyone, however young or old, was expected to play their part, colour posters depicting smiling children with catapults taking

---

[*] Unsubstantiated but widely quoted reports claim that the final total included one billion sparrows, 1.5 billion rats, 100 million kg of flies and 11 million kg of mosquitoes.

aim at the defenceless birds. Mao himself proclaimed that 'The whole people, including five-year-old children, must be mobilised to eliminate the four pests.'[3] Neither was the killing confined to China's cities: in the countryside, too, the sparrows were poisoned, trapped or caught by glue spread along the branches of trees.[4]

To encourage the slaughter, competitions were held, with rewards and praise lavished on those who could produce the greatest number of corpses. One young man from Yunnan in south-west China, sixteen-year-old Yang She-mun, became a national hero when it was revealed that he alone had killed 20,000 sparrows. He did so by locating the trees where they were nesting during the day, and then climbing them after dark to strangle the birds with his bare hands.

A cosy cottage tucked away behind a busy main road in North London's Muswell Hill is not where you might expect to find one of the very few surviving eyewitnesses to the Great Sparrow Campaign. Then again, Esther Cheo Ying has always led a somewhat unconventional existence.

A small, neat, well-dressed woman, now nearly ninety, Esther recounted the first part of her life in her painfully honest and very readable memoir, *Black Country Girl in Red China*.[5] She was born in 1932, to a Chinese father and English mother, and spent the first six years of her life in Shanghai. Then in 1938, after her parents' marriage ended, she returned to the UK with her mother and two younger siblings. Coming back to Britain wasn't the end of her childhood upheaval. As she matter-of-factly explained to me as we drank coffee together, 'My mother loved her children but couldn't look after them.'[6]   Placed into care, she and her siblings were brought up by foster parents – her beloved 'Auntie and Uncle' – in Staffordshire, in the English Midlands.

In 1949, aged just seventeen, Esther decided to return to China to join the Red Army, just as Mao's People's Republic was seizing power. What might today seem like naivety (especially as the only two Chinese words she knew were those in her name) was a burning desire to build a better world. China's communist system appeared to offer this opportunity. 'After the Second World War', she told me, 'the younger generation began to see things in a different light. We began to question things; by then I began to realise that society wasn't fair . . .' As Jill Tweedie notes in the foreword to the 1987 edition of Esther's memoir, this was also a quest to discover who she really was. 'Full of enthusiasm, intent on recapturing her Chinese identity, Cheo Ying did her best to bury the English, the "Esther" part of herself.'[7]

For a decade, she managed to do just that, though from time to time her English side would resurface, usually in the form of rebelling against the inflexibility of Mao's authoritarian regime. Even when, during a city-wide cull to prevent the spread of rabies, she was made to shoot her beloved dog, she reluctantly complied, despite being heartbroken for her loss.

Gradually, however, Esther was forced to behave in ways she found unacceptable, and could no longer avoid speaking out. 'I've always been a rebel. At the beginning I was just as fanatical as everyone. But then I started questioning things I shouldn't have been questioning.' Her rebelliousness came to a head when the order came to kill the sparrows. Esther was working at Beijing's

English-language radio station when, as she recalls, 'everything just stopped'. At dawn the next day, millions of people swarmed onto the streets, to begin the killing.

She was the only person working at the radio station who did not take part. Even her close friend and colleague Wei Ling, an intelligent and educated woman, joined in. As Esther recalls, 'she was running around like a savage, banging cymbals, with glazed eyes'.

> I sat at the window in disgust watching my colleagues, some frothing at the mouth in excitement as each exhausted bird finally fell to the ground, was stamped on and crushed to death by a shouting hysterical mob. Wei Ling triumphantly picked up one crushed bird and threw it at me, laughing.

Esther shows me a faded black-and-white photograph of her holding a baby swallow with a broken wing – one of millions of unintended casualties of the cull. 'When I found that poor little bird, that's when I really began to sabotage the campaign. I just said, "Fuck Chairman Mao – I'm not taking part!"'

I made my surprise clear: not that this redoubtable woman would think such a seditious thought, but that she would actually say it out loud, despite the potentially serious consequences. How did she get away with such a rebellion against the all-powerful God-Emperor?

Probably because, she told me, she had served in the Red Army, which gave her a certain status amongst her middle-class colleagues. Also, Esther's mother had been a charwoman before she met and married her father, which added to her working-class credentials. 'And because at that stage I couldn't speak or write Chinese fluently, I think they regarded me as a semi-literate peasant! So, I got away with murder.'

By then, it was very clear that Esther's English self had well and truly triumphed over her Chinese one.*

The details of the Four Pests Campaign reverberated around the world, with the *New York Times* – tongue very much in its cheek – calling the sparrow 'this feathered "counter-revolutionary", a menace to the five-year plan, no less'. *Time* magazine took the campaign more seriously, quoting a triumphant report from the *Peking People's Daily*, which has a striking similarity in tone to wartime propaganda against a human enemy: 'No warrior shall be withdrawn until the battle is won. All must join battle ardently and courageously; we must persevere with the doggedness of revolutionaries.'[8] According to *Time*, three million citizens took part in a single day's cull, in the Chinese capital Beijing:†

'At 5 a.m. bugles sounded, cymbals crashed, whistles trilled. The massed students beat their kitchenware and advanced, as Radio Peking recounted, singing a rousing, revolutionary anthem: "Arise, arise, oh millions with one heart; braving the enemy's gunfire, march on."'[9]

---

* Esther finally returned to Britain (via Berlin) with her two young sons in the early 1960s, eventually leaving her first husband and marrying a Jewish *Kindertransport* refugee, Lance Samson (who coincidentally also arrived in Britain in 1938, the same year as her). Only decades later, in 1980, was she finally able to return.

† Historically transcribed in English as Peking and briefly renamed Beiping when it was replaced as China's capital by Nanjing between 1928 and 1949. 'Peking' continued to be used internationally until 1979.

The most vivid contemporary account of the massacre comes from the weekly *New Yorker* magazine in October 1959, less than a year after the campaign was launched, by the Chinese-born physician and author Han Suyin. She recalled arguing with her late father's elderly servant Hsueh Mah, who was adamantly opposed to the killing of the sparrows, when their conversation was interrupted by a loudspeaker outside their house booming into life. 'Our scientists have discovered', Suyin remembers it blaring, 'that after two hours' flight a sparrow is exhausted and drops to the ground, where it can easily be caught. Our tactics in this noble struggle against this public enemy are to prevent it resting its feet anywhere – on roofs, on walls, on trees. *Keep the sparrows flying!*'[10]

The announcement left nothing to chance, going on to give detailed instructions on exactly how to achieve the campaign's aims. Housewives were instructed to tie bells to scarecrows and place them at strategic vantage-points in trees and on chimneys. Students were told to arm themselves with poles, with strips of cloth flapping at the end, and to chase and harass the birds. Others were encouraged to form squads banging household objects and shouting at the top of their voices to make sure the sparrows never had a moment's rest.[11]

The following morning, before dawn, Han Suyin witnessed the war against the sparrows for herself, as the squads of young men and women assembled in the street outside, ready for battle. At first, she recalls, the sparrows flew around in small flocks, but soon began to disperse, perching in the trees or on telegraph wires. But every time a bird tried to land, it would be dislodged with long poles or scared into flight by the constant noise and shouting. They simply had nowhere to hide.[12] Sparrows weren't the only birds disturbed by the noise: Han Suyin saw swifts, crows and magpies – 'mobs of birds, in flight, distraught, and near desperate' – flying here and there, in complete panic.

Neither, inevitably, were sparrows the only small birds to be killed. In his 2005 memoir,[13] Sheldon Lou (later a university professor in California, but at the time of the cull a teenager in Beijing) recalls voicing his misgivings to a friend: 'How can we make sure we kill only sparrows but not other birds?' Of course, they could not. Esther Cheo Ying also remembers that, for some time afterwards, she saw virtually no birds in Beijing. 'The swifts did not come back for many years. I would look up over the Gate of Heavenly Peace and the Temple of Heaven and wonder what was missing in that marvellously blue sky, and then remember: the swifts were missing.'[14]

For Han Suyin, caught up in the middle of the massacre, events then took an unexpected turn. When she returned to her father's home, she was greeted by Comrade Wong of the Street Committee, who solemnly informed her that, while the campaign was going very well, they had encountered an obstruction from her father's servant, who was refusing to allow the army of sparrow killers onto the property. Told that she must co-operate in the fight against the sparrows, Hsueh Mah countered by recalling her own, older generation's attitude of benign tolerance towards these birds. 'In my time, there was no such thing as fighting sparrows. I am a countrywoman, and we only caught sparrows to eat in time of famine.'[15]

Comrade Wong's response was short and to the point. 'There

will be no more famines *now*.' With hindsight, those words would come to have a deeply ironic significance.

The battle against the sparrows continued without respite for the whole day. The following morning, at 4.45 a.m., 'a blast of sirens catapulted us once more into the war'. Later on, Han Suyin noticed a subtle change in the surviving birds' increasingly desperate response to the continual attacks. By now, there were far fewer sparrows, and their flight was more erratic, as the surviving birds neared exhaustion. If a sparrow fell to the ground, it would immediately have a cord placed around its neck and be throttled to death, 'to join the bundles of strangled sparrows carried by sturdy boys and girls with the red scarves of Pioneers'.[16] That evening, Han Suyin saw vans laden with the corpses of tens of thousands of sparrows passing through the city streets. On their sides, hastily painted slogans pronounced 'FIGHT THE SPARROW TO THE DEATH! BRAVELY STRUGGLE FORWARD, ELIMINATE THE SPARROW PEST!' And that's exactly what had happened: in Beijing alone, an estimated 800,000 sparrows were killed in just two days.

Some birds did almost manage to escape. One flock that entered the grounds of the Polish embassy was initially given sanctuary by the diplomatic personnel, who flatly refused all requests for the Chinese mobs to enter the premises. But the birds' respite was short-lived: the embassy was soon surrounded by local people who beat their drums constantly for two whole days and nights. Afterwards, the embassy staff had to use shovels to clear hundreds of dead sparrows from the compound.

*

The Great Sparrow Campaign was, on the face of it, an unqualified success. It has been claimed that one *billion* tree sparrows were killed, and while this is likely to be an exaggeration, hundreds of millions of birds did undoubtedly perish. Soon after the cull, the species was on the brink of extinction in China. In a remarkable twist to the story, several years later 250,000 tree sparrows had to be imported from the Soviet Union to replenish the devastated Chinese population.[17]

Less than twelve months after the cull came the terrible aftermath. The rice harvest, in June and July 1959, was an unmitigated disaster. Yields plummeted, for one simple reason: although tree sparrows do feed on seeds and grains in the autumn and winter, during the breeding season they feed their hungry chicks on countless millions of insects. With all the sparrows now gone, those insects – including vast swarms of locusts, the most destructive pest of all – were able to strip the precious crops bare.[*]

Despite the growing signs of a nationwide famine, however – one which would ultimately lead to the deaths of millions of Chinese people – the killing of sparrows continued to be promoted and encouraged throughout 1959. Finally, towards the end of that year, Mao abruptly declared an end to the Great Sparrow Campaign, replacing sparrows with bed bugs. It was a colossal political U-turn, with articles in the government-run media now denouncing the cull they had so enthusiastically supported little over a year earlier.

---

[*] As the US political historian Dr Judith Shapiro points out, the Great Sparrow Campaign was not the only reason for the ensuing events: 'the resulting famine had many causes, including the steel smelting campaign and ill-advised agriculture policies such as close planting'.

Esther Cheo Ying recalls showing a copy of the *People's Daily* to her friend Wei Ling and pointing out that she had been right all along.

'I shall remember what you said about our beloved chairman,' was Wei Ling's terse retort. 'You disobeyed the order of the Party. That is unforgivable.'[18]

So what brought about this sudden change in policy? Mainly the insight and courage of two scientists, Chu Tsi and Tso-hsin Cheng, who dared to question the scientific theory that had underpinned the sparrow cull.

Tso-hsin Cheng (1906–98), also known as Zheng Zuoxin, had become interested in nature during boyhood. After studying in the United States, he returned to China to begin a career as a zoologist, specialising in birds. In 1950, a year after Mao came to power via the Chinese Communist Revolution, Cheng moved to the capital where, a year later, he founded the Peking Natural History Museum. When the Four Pests Campaign began, Cheng immediately realised the potentially disastrous implications of removing tree sparrows from the delicately balanced ecosystem. But he needed proof. Over the next year, together with his colleague Chu Tsi (also known as Zhu Xi), he methodically examined the digestive systems of sparrows.

The two scientists found exactly what they expected: during the breeding season, three-quarters of the stomach contents was made up of insects, and just one quarter was seeds and grains. The birds did take some of the crop harvest, therefore, but those losses were more than offset by the benefits from their crucial role in controlling harmful insect pests. Aware of the importance of their findings, Cheng and Tsi immediately contacted the Chinese Academy of Sciences, who in turn notified the Party, leading to that extraordinary about-turn in policy.

You might imagine that these two scientists would be hailed as heroes. But that would be to misjudge the uncompromising ideology of the Chinese Communist Party, who could never admit that it had been so utterly, and disastrously, wrong. So, although this did lead to a swift termination of the campaign, both men would suffer grievously for daring to speak out against official policy.

Tso-hsin Cheng was already under suspicion for having studied in the United States, and for his long-standing collaboration with ornithologists in East Germany and the Soviet Union. Even though he was absolutely right, daring to voice his opposition to the sparrow cull had only made the situation worse. Cheng was forced to halt his scientific work and declared a criminal, under the new anti-intellectual climate in which, in the words of one popular – and distinctly Orwellian – slogan of the time, 'The more knowledge you possess, the more you are a reactionary.' Having failed a (rigged) test purporting to examine his ornithological credentials, he was made to clean toilets as a punishment. In 1966, he was put into solitary confinement in a cowshed, while the Red Guards confiscated the thing he valued most: his precious typewriter, on which he had written so many important scientific papers.

Following the death of Mao Zedong in September 1976, the climate of hostility towards academics began to soften. Cheng's *magnum opus* on the birds of China was eventually published – though gallingly he was forced to include a foreword by Mao.[19] Gradually he was allowed more freedom, and eventually he travelled to the UK, where he collaborated with the celebrated conservationist Sir Peter Scott.

At the time of his death in June 1998, at the age of ninety-one, Tso-hsin Cheng was widely hailed as the founder of modern ornithology in China. Over the course of his long lifetime, he had published 140 scientific papers, more than 260 articles and essays, twenty monographs and thirty books.[*]

Despite this later rehabilitation, Tso-hsin Cheng never fully recovered from the fallout from the Great Sparrow Campaign. Unlike tens of millions of others, however, he did at least survive.

In terms of human suffering, the Great Sparrow Campaign was quite simply the greatest man-made disaster in the whole of human history. In less than three years, between 1959 and 1961, somewhere between 15 million and 55 million people died, in what became known, in an ironic echo of those earlier campaigns, as the Great Chinese Famine.[20] To put this in perspective, the higher end of this estimate is more deaths than the 40 million that occurred, worldwide, during the whole of the First World War.[21]

Not everyone died from hunger. As the US historian Jonathan Mirsky notes, refusing to participate in the Party's political campaigns 'could result in detention, torture, death, and the suffering of entire families'.[22] People were intimidated into keeping quiet through 'public criticism sessions', which often ended in violent attacks against any dissenters. The Dutch historian Frank Dikötter, author of *Mao's Great Famine*,[23] estimates that more than two and

---

[*] He also has two species named after him: Cheng's jird *Meriones chengi* (a type of rodent), and the Sichuan bush warbler *Locustella chengi*, discovered in 2015.

a half million people were beaten or tortured to death, while as many as three million committed suicide rather than face a slow death by starvation.

As the famine took hold, and discontent grew to the point of open rebellion, punishments for questioning the policy became even harsher. Many victims were tortured and mutilated, forced to eat excrement and drink urine; others were killed by being doused with boiling water, drowned in village ponds, or buried alive.[24] The true horror comes from eyewitness accounts, including this from Yu Dehong, the secretary of a Chinese party official from the city of Xinyang: 'I went to one village and saw a hundred corpses, then another village and another hundred corpses. No one paid attention to them. People said that dogs were eating the bodies. Not true, I said. The dogs had long ago been eaten by the people.'[25]

In his 2012 book *Tombstone: The Untold Story of Mao's Great Famine*,[26] the Chinese historian Yang Jisheng lays out the horrific consequences of the famine and its aftermath. In a single city, in the central Chinese province of Henan, one in eight people – more than one million – died. In one village, forty-four out of forty-five inhabitants perished; the only survivor, an elderly woman, went insane. Elsewhere, an orphaned teenage girl killed and ate her four-year-old brother.

So desperate did people become that when someone in a household died, the rest of the family would keep their body and pretend they were still alive, so they could continue to collect their dead relative's meagre food ration. Meanwhile, the rotting corpse was being eaten by mice – ironically, another of the 'four pests' that was supposed to have been eradicated by now.

Yang, who was in his late teens when these terrible events occurred, recalls coming home from school to find that his father was dying. 'He tried to extend his hand to greet me but couldn't lift it . . . I was shocked with the realisation that "skin and bones" referred to something so horrible and cruel.'[27]

Yet so strong was Yang's belief in the Communist Party that he did not connect his own family's loss with that of millions of others throughout the country. As he recalls, he believed his father's death was unconnected to the wider political and social consequences of government policy, but rather a purely personal misfortune.[28]

The Great Sparrow Campaign did not appear out of nowhere. It was the culmination of a long series of errors in both policy and practice during the previous decade, fuelled by a lethal blend of scientific illiteracy and unbridled power. The blame can clearly be directed at one man, arguably the most omnipotent individual in the history of the world: Mao Zedong.

Mao's uncompromisingly hardline political ideology, known as Maoism, has been described as 'one of the most ambitious attempts at human manipulation in history'.[29] It was, without question, the most effective way of turning 650 million people into a single, coherent nation, with all citizens expected to show total obedience to Mao's God-like rule.* But this ruthless and single-minded approach involved not just oppressing his own people, but also waging a continual war against nature.[30] As the environmental activist Dai Qing

---

* Even Esther Cheo Ying contends that this was probably the only method of governance that would have worked in a nation where over 90 per cent of the people were illiterate.

points out, Mao was that most dangerous of beings, an environmentally and ecologically illiterate tyrant: 'Mao knew nothing about
animals. He didn't want to discuss his plan or listen to experts. He
just decided that the "four pests" should be killed.'[31]

Although it may be the most serious and deadly example, the
Great Sparrow Campaign is far from the only time that human
hubris and ignorance have led to an all-out war against birds. When
humanity wages war on birds – despite our superior knowledge,
technology and resources – we often end up on the losing side.
Nowhere is that more evident than in a bizarre conflict that took
place almost a century ago, in the deserts of Western Australia.

Soon after daybreak on a fine spring morning in November
1932, the battle began. Vehicles carrying troops from the Royal
Australian Artillery reached the Western Australian settlement of
Campion, built to house former soldiers who had fought in the
First World War. The men rapidly set to work unpacking two Lewis
light machine guns and 10,000 rounds of ammunition. Within
minutes, they had sight of the enemy: fifty of them, each almost
two metres tall, gathered together on the deep red, iron-rich soil
that dominates this arid landscape.

Major G. P. W. Meredith, commander of the Seventh Heavy
Battery, ordered his men to encircle the enemy and drive them
towards the guns. Unfortunately, he had underestimated his foe:
they split up into smaller groups and, though the soldiers did
manage to shoot and kill a dozen or so, the rest escaped.

Two days later, Meredith and his men set up an ambush, and
spotted no fewer than 1,000 individuals heading their way. Surely

now they would succeed in their mission. But luck was not on their side: after just a few shots had been fired, the Lewis gun jammed and the approaching group fled to safety.

For the next few days, Meredith and his troops continued to stalk their quarry. In a change of tactics, he mounted one of the machine guns onto a truck, but their opponents could still outrun them, and the drive was so bumpy that the soldiers were unable to fire the gun. One soldier ruefully noted that the enemy were becoming more organised as the days went on: 'Each pack seems to have its own leader now – which stands fully six feet high and keeps watch while his mates carry out their war of destruction and warns them of our approach.'[32]

By 8 November, almost a week after the campaign had begun, the troops had fired 2,500 rounds of ammunition, with very little success. As the realisation dawned that they had grossly misjudged their adversaries, orders were given to withdraw. A rueful Major Meredith compared them to another legendary cohort of warriors: 'They can face machine guns with the invulnerability of tanks. They are like Zulus whom even dum-dum bullets could not stop.'[33] The only consolation was that, as his official report on the affair duly noted, the soldiers had not suffered any casualties themselves. Which was hardly surprising, given that the enemy was not human, but Australia's largest bird: the emu.

This initial campaign marked the opening salvo of what would become known as 'the Great Emu War'. Like the story of the tree sparrow, it is another salutary tale of the perils of trying to enforce a form of 'ethnic cleansing' against a single species of bird.

The war had come about for what, at the time, seemed a very good reason. Emus had always wandered into this area of Western Australia from the vast, arid area known as the Outback, in search of food and water. But now that settlements were being built here and the land was being farmed, the emus had become a major problem. The issue came to a head in October 1932, as a result of the world-wide Great Depression, which reduced wheat prices, leading to major financial hardship for the farmers. That was when, in a classic case of bad timing, more than 20,000 emus arrived on the scene.

Emus are one of the tallest and heaviest birds in the world,[*] which meant these birds were serious adversaries. They didn't just eat the crops, but trampled them, too, and also knocked holes in fences, allowing another 'pest species', the rabbit, to enter the fields and feed.

Something clearly had to be done. A delegation of former soldiers went to lobby the Minister of Defence, Sir George Pearce, with the suggestion that machine guns be used to kill and disperse the flocks of emus. He agreed straight away: not only would this provide much-needed target practice for the army, but it would also pay off politically, staving off rebellion by Western Australia's rural community which might result in the state gaining independence from the national government.

Assuming that the troops would be triumphant, Pearce had arranged for a cameraman from the news outlet Fox Movietone to attend, to record the event for the cinema newsreels. The resulting

---

[*] A fully grown emu *Dromaius novaehollandiae* can reach a height of 1.9 metres, weigh as much as 60kg and run at speeds of 50 kph over short distances – easily enough to outpace a human pursuer. Only ostriches are taller, while they and cassowaries are also heavier.

short film is a classic example of the power of propaganda over an inconvenient truth.[34]

The film opens with typically jaunty music and the caption 'WESTERN AUSTRALIA MAKES WAR ON EMUS – Army machine guns are called in to help farmers at Campion repel mobs of marauding birds.' The newsreader then sets the scene in his clipped tones, before delivering a jokey commentary referring to 'the scouts of the advancing army', 'our lads' and 'the enemy watching events through their periscopes' – the emus' long, extended necks. He concludes with this optimistic – and as it turned out, totally false – statement:

'Instead of the birds ruining the farmers, it seems the tables are turned, and there'll be no more damage done here for many a day to come.'

Less than a week later, with the emus continuing to devastate the precious crops, there was a second attempt at a cull. Again, the numbers of birds killed were pitiful: about a hundred a week, at which rate it would take years to have any real effect. It is also likely that Meredith exaggerated the total of emus killed in order to save face.

By then it was too late, as the debacle was already being debated in parliament. When one parliamentarian was asked if an official medal should be issued to the soldiers involved, he responded sourly that if medals were to be given out, they should go to the emus, as they had 'won every round so far'.[35]

To this day, the Great Emu War remains the only known example in history of an official army being defeated by a bird. Or, as we might say, Emus 1, Humans 0.

*

Back in China, more than sixty years since the Great Sparrow Campaign was launched, one key question remains: what long-term effects, if any, did the cull have on the fortunes of the species at its centre?[*]

Songbirds such as the tree sparrow have evolved to be able to recover rapidly from unusual events, be they natural, such as a hurricane or a harsh winter, or man-made, like the sparrow cull. They have two or three broods of chicks each year, and can produce ten or more youngsters in a single breeding season. This means that, so long as their habitat and food supply remain intact, numbers can very rapidly return to normal. That is evidently what happened in China.[36]

However, worldwide, the tree sparrow population is showing a slow but steady decline, though not yet at a fast enough rate to cause concern. Local falls can, of course, be far more serious than global ones: in the UK the tree sparrow is one of the most rapidly declining birds, numbers having dropped by about 95 per cent since the 1970s, mostly because of habitat loss and a shortage of suitable food, though recently the population does appear to have stabilised.[37] In Europe, too, tree sparrows are steadily declining.

The species is still, however, very common and widespread throughout the temperate regions of Asia. In China itself, population figures are hard to come by, but sparrows are a common

---

[*] The good news is that, according to BirdLife International, the Eurasian tree sparrow is currently in the category of 'Least Concern'. The global population is estimated at between 190 and 310 million individuals, and the breeding range covers roughly 99 million square kilometres. See BirdLife International website.

breeder in many urban and rural areas. In China's megacities, though, they do appear to be under threat; not from being deliberately killed, but from development, habitat fragmentation and human disturbance.

In 2008, three Chinese researchers published a case study which looked at tree sparrows in China's second largest city, Beijing, during both the breeding season and the winter.[38] With over 21 million inhabitants, Beijing is now more than seven times as populous as it was in 1958, when the sparrow cull took place. Growing at an annual rate of more than 2 per cent, the city's human population is predicted to reach over 25 million by 2037.

This is bad news for Beijing's tree sparrows. By studying eight different urban areas, ranging from suburbs and parks to the city centre, the researchers discovered that the more urbanised an area becomes, with a higher density of buildings and roads, the greater the decline in sparrows. Conversely, the presence of trees, and green areas such as parks, allows the population to maintain its former levels. The study concluded that 'the tree sparrow had not adapted to rapid urbanisation *even though it is a generally adaptable species* [my italics]', and also that 'urban planning should take birds . . . into consideration.'

Meanwhile, in December 1993, a consignment of two million frozen tree sparrows was discovered at the Dutch port of Rotterdam. They were in transit from China to Italy, where presumably they would have been destined for human consumption.[39] As the late ornithologist Denis Summers-Smith – the world expert on sparrows – noted, 'Although there is nothing illegal in this trade, cropping at this rate must surely be more than the population can withstand.'[40]

In Hong Kong, the first ever tree sparrow census was carried out in 2016.[41] Sparrows are one of the few birds that thrive in Hong Kong's busy centre, and the census confirmed this, with an estimated total of 320,000 birds. The survey has been carried out every year since: the results from 2020, estimated a total of 260,000 sparrows – an average of 235 per square km.[42]

On a global scale, the tree sparrow's close relative, the house sparrow, continues to benefit from its long-standing habit of living 'almost exclusively in complete dependency with man'.[43] Thanks to this intimate relationship with human beings, the house sparrow has been able to spread out from its native range in Europe and Asia, to colonise North and South America, Africa and Australasia, and is now one of the commonest and most successful birds in the world. However, during the past few decades, house sparrow numbers in Europe have fallen by almost 250 million, showing that, however common and widespread a bird may be, we cannot, and must not, take its continued presence for granted.[44]

What wider lessons can we learn from the events of 1958? In China, it seems, very few. Today, more than six decades after that enormous human death toll, the Great Chinese Famine remains taboo. When mentioned at all, the famine is usually described euphemistically as the Three Years of Difficulties, or the Three Years of Natural Disasters, with historians suggesting that extreme weather events, such as floods and droughts, were to blame. This has been strongly refuted by Frank Dikötter, who has demonstrated that the famine not only lasted longer than had previously been thought – for a full four years – but that its causes were also almost entirely political.

'The term famine', he concludes, 'brings to mind the absence of food and people somehow slowly starving to death. It's a very passive sort of image . . . A better term would be "mass murder".'[45]

The celebrated Chinese-British author Jung Chang, and her husband, the Irish historian Jon Halliday, would no doubt agree with that verdict. In their monumental 2005 biography of Mao Zedong, they reveal the truth: that Mao was prepared to accept that mass deaths would be the inevitable consequence of the draconian policies enacted in the Great Leap Forward.[46] 'This is the price we have to pay,' China's Foreign Minister Chen Yi announced in November 1958 once the rising death toll became clear; 'it's nothing to be afraid of. Who knows how many people have been sacrificed on the battlefields and in the prisons . . .? Now we have a few cases of illness and death: it's nothing!'[47]

That was a view shared – indeed, probably originating from – Mao himself, who a year later, at a secret meeting in Shanghai, proclaimed: 'When there is not enough to eat people starve to death. It is better to let half of the people die so that the other half can eat their fill.'[48]

The historian Rana Mitter, reviewing Yang Jisheng's book *Tombstone*, suggests that all famines are essentially the result of political decisions rather than natural disasters. As he points out, even when the disastrous consequences of China's misguided policies became clear, the politicians refused to change course:

That was the moment at which the leadership lurched into criminal irresponsibility . . . Yang Jisheng's book is not just a tombstone for his father and other famine victims, but for the reputation of the Communist party's leadership at a time when they should have acted – and failed to do so.[49]

Just before Esther Cheo Ying and I said goodbye, I asked if she had ever confronted her former colleagues over what had really happened during the sparrow cull all those years ago. She had, she said. 'You were right,' was the response from one of them. 'But at the wrong time.' As a verdict on the futility of dissent under the yoke of a totalitarian dictatorship, it's hard to beat.

The Great Sparrow Campaign, and the unprecedented human disaster to which it contributed and partly caused, shows the danger of a society losing touch with fundamental ecological truths. As the British environmental writer Michael McCarthy noted in 2011:

> Of all the egregious acts of dictators, down through the centuries, Mao's sparrow death sentence is in a weird class by itself, not just for its scale, or its ruthlessness, but because in his own terms it was rational, a logical consequence of the pursuit of scientific socialism, and he possessed the means to carry it out – 600 million people who would obey his every whim – yet it was delusional entirely.[50]

For McCarthy, Mao's campaign against the sparrows typifies the Chinese leader's dangerous and misguided attitudes towards the natural world: nothing less than the delusion that he could actually conquer nature – submit it to his own wishes and desires, just as he had the Chinese people. 'Nature was not there to be respected;

it was simply a resource to be used, and the mountains and rivers were to be mastered; they were to be bent to man's will.' Mao's slogan – at the very heart of his political and economic philosophy – was 'People Will Conquer Nature.'[51]

Just how out of kilter such a philosophy now is with the outlook of a great many Chinese citizens, is clear from this satirical self-interrogation by the playwright Sha Yexin in 1997, almost forty years after he had witnessed the events for himself.

> Sparrows have flaws just like intellectuals. When they jump on the flagstaff, they look self-important. They also like to chirp when they are in charge, just like intellectuals like to talk . . . But ultimately sparrows can catch harmful insects and fulfil their functions, just as intellectuals can toil honestly and do what they are good at . . . their contributions are greater than their flaws . . . So how could we butcher them? There may be a few misbehaving individuals, but that is just a very small number for which there is no reason to mobilise the entire human population to eradicate the entire species.[52]

Or, as the nature writer (and Esther Cheo Ying's grandson) Charlie Gilmour remarks, Mao really was the totalitarian version of the old lady who swallowed a fly.[53]

But perhaps the biggest lesson from the Four Pests Campaign – and the manner in which politicians and people jointly embraced, with alacrity, such an environmentally disastrous policy – is that it could so easily happen again.

Indeed, in China, it already has. In 1998, forty years after the initial Four Pests Campaign was launched, it was temporarily revived in the south-western city of Chongqing. Posters appeared exhorting people to 'Get rid of the Four Pests.'[54] Then, in January 2004, the Chinese government announced that it was launching a Maoist-style 'patriotic extermination campaign' against civet cats, badgers, racoon dogs and cockroaches, which were accused of harbouring and spreading the potentially lethal Severe Acute Respiratory Syndrome (SARS) virus.[55] As the World Health Organisation (WHO) was quick to point out, though, the cull could well prove counter-productive. Coming into close contact with the animals while killing and disposing of them could make people *more* likely to catch, and then inadvertently spread, a dangerous disease.

Not only is China unable to learn the lessons of its past mistakes, it also seems fated to repeat them. But this is far from the only example of all-powerful leaders adopting knee-jerk, populist policies which then lead to potentially irreversible damage to both people and the environment:[*] take former president Trump of the USA and Jair Bolsonaro of Brazil. Add in that our generation now faces the greatest existential threat to humanity, wildlife and the planet: the unprecedented climate emergency.

In the final chapter, I shall focus on one species that, above all others, embodies the collateral damage of that threat.

---

[*] Or *not* adopting policies, as when China failed to prevent the spread of the Covid-19 virus via 'wet markets', where live wild animals were sold for food.

10

EMPEROR PENGUIN

Aptenodytes forsteri

Great God! This is an awful place . . .

Robert Falcon Scott, *Journal*, 17 January 1912

Of all the many dramatic scenes the film crew had witnessed during their long stay in Antarctica, this was by far the most distressing. A group of fifty emperor penguins, most carrying a single chick balanced precariously on their feet, had somehow slipped down the steep sides of a gully. Now they were unable to escape.

This presented the production team with a tricky ethical dilemma. If they did nothing, the penguins would eventually starve to death. But if they decided to intervene, and help the birds reach safety, they would transgress the code of conduct for wildlife filmmakers, whose most sacred principle is 'never intervene'.

This might sound cruel, but the code is there for a purpose. As soon as a filmmaker intervenes to influence a natural process – for example, by saving a helpless baby antelope from being killed by a lion – they have stepped over that invisible boundary separating the observer from the observed and interfered with nature. As the sonorous, calming voice of Sir David Attenborough explained, over shots including a chick that had already frozen to death, 'Film crews have to capture events as they unfold, whatever their feelings.'

It seemed that the team, who were in Antarctica to film the emperor penguin colony for the BBC series Dynasties,[1] *could do nothing to help. This did not, of course, stop them from caring deeply about the terrible calamity they were being forced to witness. Cameraman Lindsay*

McCrae wiped a tear from his eye as he spoke: 'I know it's natural . . .
but it's bloody hard to watch.'

Soon afterwards, however, there was a glimmer of hope. One plucky
penguin, still carrying a chick on its feet, managed to clamber out of
the gully. The crew watched as, using its short but powerful wings, and
finally its beak, it laboriously levered its way to safety.

Two days later, when the weather had cleared and they could return
to the gully, they were heartbroken to see that the situation had deterior-
ated. Knowing that they could not intervene directly, for example by
physically carrying the penguins out, they nevertheless decided to give the
birds another option. So they dug a ramp, with a series of shallow steps
cut into the ice, to give the birds a chance to escape. To their relief, the
penguins took advantage of the new route out of the gully. All managed
to survive.

When the episode of Dynasties *featuring the penguins was broad-*
cast, the BBC made the bold decision to place the rescue at the centre
of the story. In the publicity for the series, they also made it clear
exactly how and why the film crew had decided, despite that rigor-
ous code of conduct, to intervene. To their delight – and possibly
surprise – the response from both conventional and social media was
overwhelmingly positive. The crew's decision had, it seemed, been
vindicated.

Soon afterwards, the director Will Lawson – billed as 'the Man Who
Helped Save the Penguins' – was interviewed by Lorraine Kelly on her
eponymous ITV morning show. At one point, Kelly noted that 'David
Attenborough said it was all right for you to help,' thus inadvertently
casting this much-loved national treasure in an even higher, God-like
role.

'In some respects', Will Lawson reflected, 'the environment is almost like a predator.'

Lawson had hit the nail squarely on the head. For the main threat facing not just emperor penguins, but all the world's living creatures is environmental change: more specifically, the climate crisis. *

Emperor penguins have evolved to do what no other living creature can: breed and raise their single chick slap-bang in the middle of the impossibly cold and harsh Antarctic winter.

To do so, the species has, over many millennia, developed one of the most complex and challenging breeding cycles of any of the world's 10,000-plus species of birds. And the strategy has worked: in 2009, a survey of the emperor penguin's forty-six known colonies from space – the first ever attempt to map an entire species' numbers – estimated a global population of roughly 600,000 adult birds, double the number previously assumed.[2]

At that point, and indeed until a decade or so ago, the species was judged by the global conservation organisation BirdLife International as being of 'least concern'.[3] But in 2012, after further research on the species' recent decline, the emperor penguin was moved up to the 'near threatened' category.[4] Yet, unlike most declining birds, which are usually threatened by a combination of habitat loss, persecution, pollution and other relatively local-ised causes, the emperor penguin is facing oblivion for one reason alone: the global climate crisis.

* Following recent shifts in the terminology used by scientists and parts of the media, I do not refer to 'global warming' (which sounds rather benign), or 'climate change' (which sounds neutral), but the more urgent 'climate crisis'.

It is not the only species that will struggle to survive if we do nothing to prevent the world's climate heating up so rapidly. But it is, like the polar bear at the opposite end of the planet, one of those we notice and care about the most: the 'poster bird' of the impending catastrophe.

Although the emperor penguin is likely to reach the tipping point towards inevitable extinction before many other species, we know that thousands of others will soon follow; currently, one in eight of the world's birds is under threat.[5] While we may regret this loss of biodiversity, in terms of the extinction of species, and the loss of the places where they live, we should also be aware that another species – *Homo sapiens* – is also facing an existential crisis. We might arguably be the most successful species ever to live on our planet, but that will not save us from the devastating consequences of climate chaos.

Ironically, the much shared, and deservedly praised, YouTube clip of the film crew rescuing the penguins is the perfect metaphor for the wider climate crisis. While they – with us viewers rooting for them – focus on the immediate plight of those individual birds and do their very best to save them from danger, we continue to ignore the bigger picture of the unprecedented threat facing us all.* It is as if we simply cannot cope with the sheer enormity of what we now face.

---

* The late Miles Barton, producer of the *Dynasties* 'Emperor Penguin' episode and a much-missed colleague and friend, did conclude the programme with a timely warning about the potentially catastrophic effects of the climate crisis on the penguins: 'The annual disappearance of their ice-world is a reminder that they face an uncertain future. Ocean temperatures are expected to rise year-on-year. This Antarctic sea ice, on which all emperor penguins rely, may not freeze long enough each year for them to complete their extraordinary life cycle.'

Those particular penguins – along with their very cute and fluffy chicks – were spared a slow, icy death and, thanks to human intervention, survived to create a new generation. But their descendants will dwindle and decline until one day – probably in our grandchildren's lifetimes – the world's largest species of penguin will, like the dodo, go extinct. By then, it is quite possible that humanity will also have crossed the tipping point between survival and our own, apocalyptic, self-destruction.

That is the theme of this final chapter: can we – and the birds – survive?[*]

The world's birds can broadly be divided into two categories: generalists, which thrive in a wide range of habitats and exploit many different ecological niches; and specialists, which have evolved to suit a very specific niche.

Of the ten species or groups in this book, four are generalists (raven, pigeon, bald eagle and tree sparrow), two are somewhere in between (wild turkey and snowy egret), while the remaining four are out-and-out specialists (dodo, Darwin's finches, guanay cormorant and emperor penguin). Of this quartet, the emperor penguin has evolved the most complex and specialised life cycle; one that might seem bizarre to us on first impression, but to this species makes perfect sense.

No other species of bird – nor indeed, any other living creature

---

[*] I have chosen not to examine the specific causes of the current climate crisis; these – and especially the role of man-made (or 'anthropogenic') climate change – are now widely accepted. See the 2006 Stern Review: 'The Economics of Climate Change', and many other subsequent reports.

– spends virtually the whole of its life in the interior of Antarctica, let alone chooses to breed there. For the emperor penguin, this involves braving temperatures which commonly drop below 40 degrees Celsius with added severe wind chill, and living for one-third of the year in darkness. This raises several questions, of which the two most obvious are: why would they do so, and, equally importantly, how on earth do they manage to survive?

'Why' is easy to answer: by adapting to living there, and by breeding in large, dense colonies numbering several thousand birds, they reduce the threat from predators, especially during the early part of the breeding season, when first the eggs, and then the chicks, are at their most vulnerable. That is not to say that the colonies are entirely safe. Southern giant petrels – the largest members of their family, the size of a small albatross – scavenge dead and dying adults and chicks, and also actively predate the youngsters.[6] Adult penguins may also fall victim to two predatory marine mammals, the orca and the leopard seal, when they return to the open ocean to feed.[7] On the ice, however, they are more or less safe.

Size – and weight – are crucial to the emperor penguin's success. As its name suggests, this is the largest of the world's penguin species, and the world's heaviest seabird. Indeed, in terms of weight, it is the sixth heaviest bird on Earth,[*] weighing between 22 and 45kg. Males and females, which look alike, stand roughly 100cm tall.

Like other flightless birds, the emperor penguin has been able to achieve its height and weight by dispensing with the power

---

[*] The top five are all flightless species of ratite: the African and Somali ostriches, southern and northern cassowaries, and the emu.

of flight, and so no longer needs to prioritise lightness over the many advantages of being large and heavy. The name of its genus, *Aptenodytes*, translates from the Greek as 'diver without wings'. Like all penguins, the emperor's short and stubby, yet incredibly powerful, flippers (which did originally evolve from wings) enable it to dive to extraordinary depths, and at great speeds, in pursuit of its prey.*

Size and weight also enable the emperor penguin to survive during the harsh Antarctic winter. As the 'square–cube law' notes, as a shape increases in size, so the ratio of its surface area to volume decreases. When applied to living organisms, this simply means that the larger the creature is, the better it is able to retain heat compared to a smaller creature of the same shape. Thus, a large penguin loses heat more slowly than a small one.

Paradoxically, this can cause problems for the emperor penguin, even in this coldest of environments. In cold weather – and especially high winds – emperor penguins enter a 'huddle', in which those in the centre are able to stay warmer than those on the edge, as the temperature there sometimes reaches an incredible 20–30 degrees Celsius above freezing. But watch a huddle for any length of time, and it soon becomes clear that the birds are constantly changing their position, so that those that were on the inside are now on the periphery, and vice versa. The traditional explanation, repeated in many books, articles and TV programmes, is that

---

* According to Guinness World Records, the deepest verified dive of any bird was made by an emperor penguin: one bird, fitted with a tracking device, reached a depth of 564 metres off eastern Antarctica. In contrast, the deepest recorded dive by any flying bird is just 210 metres, by a Brünnich's guillemot (thick-billed murre) – well below half the depth achieved by the emperor.

the penguins are acting selflessly in order to allow their peers to take their own turn in the middle, and so ensure the survival of the colony. It's a superficially convincing theory, but it is, in fact, wrong. Emperor penguins are so well insulated, with a thick layer of fat beneath those densely packed feathers, that if they spend too long in the centre of the huddle they are in serious danger of overheating. This is why they move.[8]

The complex breeding cycle of the emperor penguin has featured in many popular wildlife television documentaries.[9] The story is always the same, though the drama loses little in the retelling. At the beginning of the freezing Antarctic winter, in March or April, both male and female penguins travel up to 120km inland from the edge of the pack ice. Because they are unable to fly, they must make this long and arduous journey on foot.

Once the birds have reached the site of the breeding colony, the male starts to display to the female, who responds by mirroring his movements and sounds, to cement their pair bond. Despite their longevity – emperors can live for an average of twenty years, and sometimes far longer – they do not actually pair for life, but do stay faithful for the whole of the breeding season.

After mating, the female lays a single, pear-shaped egg weighing roughly 460 grams,* which she almost immediately transfers to the top of her mate's feet. If she drops the egg onto the ice, or he fails to receive it properly, then unless they can get it safely back onto the male's feet within the next minute or so, the chick inside is

* Despite its large size – about seven or eight times the weight of a chicken's egg – this is only just over 2 per cent of her body weight, meaning that it is one of the smallest eggs in the world relative to the bird's size.

doomed. Miss this brief window and the entire breeding season has gone to waste, almost as soon as it began.[10] However, if the egg is transferred successfully, then the female almost immediately departs, heading back to the ocean to feed and rebuild her depleted physical resources. She will not see her mate again for many months.

Now the male penguin begins his long and lonely vigil. For more than two months, his only duty is to incubate the precious egg balanced on top of his feet. He keeps it snug and warm using a bare brood patch on his lower belly, which transfers the heat from his body to the chick inside, allowing it to develop and grow. When the egg finally hatches in July or August – the depths of the Antarctic winter – the male will not have eaten for four months, during which time his weight will drop by more than half, to as low as 18kg.

Even after it has hatched, the chick is still incredibly vulnerable, and entirely dependent on the male for warmth and food: a protein- and fat-rich substance known as 'crop milk'. After a few days, this will run out; if the female has not returned from the ocean by then, the chick will starve to death. If she has, then, in another risky and delicate manoeuvre, the male transfers the chick from his feet to hers; she then regurgitates half-digested food – mainly fish, krill and squid – from her stomach. The male bids her farewell, and then he too heads back to the ocean to feed, and replenish his own lost weight.

Once the chicks are six or seven weeks old, they can safely be left alone, as both parents make the long journey to the ocean and back to collect food. During this time, the chicks gather into crèches

– which keeps them both warm and relatively safe from predators such as the giant petrels or South Polar skuas (imagine a gull on steroids). Soon afterwards, the youngsters begin to moult from their downy covering into adult plumage. By the time summer arrives, in December and January, they make that long journey to the ocean, where they must now survive on their own.

The emperor penguin's breeding cycle does appear to be very complex and risky – and in many ways it is. But it works – or at least it did until recently.

Now, however, because of rapid changes in the climate of Antarctica, the bird's entire life cycle is at risk of falling apart. This is down to the rapid and unprecedented changes to the sea ice on which the penguins depend, and the knock-on effects of these changes, both to their breeding habitat and food supplies.

Put simply, as the extent of the sea ice reduces, the penguins in those colonies around its edge will not be able to complete their breeding cycle quickly enough before the ice begins to break up during the spring and summer, so their chicks will drown. This has already happened: in 2016, when sea ice levels were disastrously low, 10,000 emperor penguin chicks drowned at a colony in Halley Bay – until then, the second largest in the world.[11]

In a further devastating consequence of ice melt, the food resources on which these birds depend – huge shoals of fish, squid and krill – will either decline or move away from their present locations. If that happens, the adult penguins will not be able to get enough food for them, and their chicks, to survive. And this does not even begin to consider other consequences of the reduction of sea ice, such as the expected catastrophic rise in sea levels, changes in patterns of wind and precipitation, and extreme weather events, all of which are likely to have a negative – and potentially terminal – outcome for the emperor colonies.

Predicting future changes in the extent of sea ice across the vast Antarctic continent, and what effects these changes will have on the emperor penguin, is not an exact science. But using new computing modelling techniques, scientists have now come up with a broad range of likely scenarios.

In 2014, a peer-reviewed paper in the journal *Nature* analysed likely trends for the entire world population of emperor penguins.[12] If trends in the reduction of sea ice continued as the models have predicted, the authors concluded, then by 2100 all colonies would have declined, two-thirds of them by more than half.

Seven years later, by 2021, the forecast had become even worse. This time, the scientists predicted that 98 per cent of all emperor penguin colonies would become 'quasi-extinct' by the end of this century. Some birds might possibly manage to find new places to feed and breed, but even the most optimistic assessment was that more than four-fifths of all the world's emperor penguins will disappear.[13] Even in the shorter term – within a single human generation – the prognosis is grim. By 2050, less than three decades into the future, seven out of ten colonies are predicted to become quasi-extinct.[14]

What this means in practice is that, although individual birds will continue to survive, in the longer term the colonies are no longer viable. In effect, the emperor penguin will become a 'zombie species' – still alive, but ultimately doomed. It is an ironically appropriate fate for a bird whose history alongside human beings has been a brief, but often troubled, one.

Given its remote and inaccessible home, it is hardly surprising that, despite its size and former abundance, the emperor penguin was not even described to science until the mid-nineteenth century.*

Little was known about the species for a further fifty years or so. Then, in 1901, Edward Wilson, the assistant surgeon (and ornithologist) on Robert Falcon Scott's expedition ship *Discovery*, came across the first known breeding colony. This was beneath the cliffs at Cape Crozier, on Ross Island, at a latitude of more than 77 degrees south.

Wilson first proposed the idea that the chicks he observed must have hatched from their eggs in the middle of the Antarctic winter – a behaviour he described as 'eccentric to a degree rarely met with, even in Ornithology'.[15] His findings caused a sensation in scientific circles, with widespread speculation that the emperor penguin represented one of the most primitive of all avian life forms.

Nine years later, in 1910, Wilson returned with Captain Scott on the *Terra Nova* expedition which, in its failed attempt to be the first to reach the South Pole (ahead of the Norwegian explorer Roald Amundsen), would go down as one of the greatest disasters in the history of exploration. But before the team could make their

---

* By the English zoologist George Robert Gray, in his 1844 work *Genera of Birds*. He gave the specific name *forsteri* in honour of the German naturalist Johann Reinhold Forster, who accompanied Captain Cook on one of his transglobal voyages, and was probably the first person ever to set eyes on the species.

ill-fated journey to the pole, Wilson and his colleagues had important work to do. They wanted to test the scientific theory known as 'recapitulation', which suggested that you could work out the evolutionary development of a species – for example whether or not a 'primitive' bird like the penguin might have evolved from reptiles – by closely examining the development of its embryo.

To do this, they needed to obtain specimens of the emperor penguin's eggs. So, in the middle of the long, dark Antarctic winter of 1911, three men – Edward Wilson, Apsley Cherry-Garrard and Henry 'Birdie' Bowers – left the comparative safety and comfort of the expedition's base camp, to trek the 100 kilometres across the ice to the breeding colony at Cape Crozier.

It is hard to imagine what a terrible ordeal this must have been for those men. Fortunately, we do not need to, because Cherry-Garrard – the only one of the trio to survive the main expedition – described it in his aptly titled book *The Worst Journey in the World*.[16] In its pages, the author describes the horrors of the five-week-long journey in plain, unvarnished detail: the darkness, biting winds, driving snow, and temperatures plummeting as low as minus 60 degrees Celsius. As he wrote, with classic British self-restraint, 'Antarctic exploration is seldom as bad as you imagine, seldom as bad as it sounds. But this journey had beggared our language: no words could express its horror.'[17]

When the men finally reached the colony, to their great disappointment they found just a hundred adult penguins present, huddling together for warmth beneath the cliff. Nevertheless, they did manage to obtain five eggs, which they wrapped inside their woollen mittens to keep warm. On their even more arduous return journey, two of the eggs broke; but, once safely at base camp, after receiving a heroes' welcome from Scott and the rest of the expedition members, Wilson managed to remove the three remaining embryos, which he pickled to preserve them.

Once Cherry-Garrard had finally returned home – following the tragic deaths of Scott, Wilson, Bowers, Evans and the fifth member of the polar team, Lawrence 'I am just going outside, and may be some time' Oates – he duly delivered the remaining emperor penguin embryos to the Natural History Museum in London. Not only was he kept waiting for several hours but, after the museum curators reluctantly took the specimens, they remained unseen, in storage, for a further two decades.[18] When they were finally examined, they were too well developed to be of any use; besides, by then, the theory of recapitulation had been soundly rejected. Those brave men's dreadful journey had, it seems, been entirely in vain.

The 2021 paper that revealed the disastrous effects of sea ice losses on the emperor penguin begins with an unusual call to arms, reflected in its title: 'The call of the emperor penguin: Legal responses to species threatened by climate change'.[19]

The authors suggested that one potential way of highlighting the birds' plight, and at the same time forcing governments (notably that of the United States) to try to mitigate future rises in temperature, would be to add the emperor penguin to the list of species protected by the Endangered Species Act (ESA). This suggestion was immediately taken up by the US Fish and Wildlife Service, who on 4 August 2021 – just a day after the paper was published – proposed doing just that.[20]

Currently, only a handful of other species are listed under the US's ESA, including the polar bear and bearded seal. The focus on these particular species may be because climate change is being shown to be happening much faster – and with more serious consequences – at the poles than at the equator, or in temperate and tropical regions.[21] But while it is clearly a good idea, listing the emperor penguin still fails to recognise one crucial aspect of the life cycles of many other birds: they migrate. Those species that head north each spring, to breed in the Arctic regions, are likely to be the next major casualties of climate change.

Take a wading bird that is very familiar to birders on both sides of the Atlantic: the red knot. Plump and short-legged – it is often likened to the shape of a rugby ball or American football – but with long, pointed wings, the knot has evolved to make epic twice-annual journeys across the globe. Some travel as far as 14,000 km each way, from the high Arctic to Tierra del Fuego. Over a single knot's lifetime, it can fly as far as the Moon and back – as was calculated for one long-lived individual, who had reached the age of at least twenty years old when last sighted in May 2014. Officially known as 'B95', this globetrotting wader was given the nickname 'Moonbird'.[22]

Tightly packed feeding flocks of these migratory shorebirds can often be seen along the coastlines of North and South America, Europe, Asia, Africa and Australasia. The species has not yet been officially sighted in the emperor penguin's home continent of Antarctica, but that must surely only be a matter of time.

With a huge world range – ranging from 50 degrees north to 58 degrees south, and covering 18 million square kilometres – the red knot could hardly be more different from the emperor penguin, whose range is tiny by comparison. Yet the complex nature of the red knot's life cycle makes it equally vulnerable to our changing climate. As a result, BirdLife International has now classified the species as 'near threatened',[23] while in 2014 the US Fish and Wildlife Service reclassified the species as being under threat, under the Endangered Species Act.[24]

The first problem facing the red knot is what is happening in its breeding range. Knots nest on tundra, usually near the coast, and virtually all breed within the Arctic Circle: in Alaska, Canada, Greenland and Siberia. Here, as we have noted, temperatures are rising far higher – and at a much faster rate – than elsewhere on the globe. In March 2022, about the time of the Spring Equinox, parts of the Arctic were experiencing temperatures more than 30 degrees Celsius above average; less than a year earlier, parts of Canada also experienced summer temperatures that did not just break previous highs but shattered them.[25]

Like all long-distance migrants, the knot times its return to its breeding grounds very precisely, using tiny changes in the relative amount of daylight. These trigger signals in its brain to begin that long journey north. Unlike changes in temperature, shifts in the amount of daylight are constant, which means that knots set off on their journey at more or less exactly the same time each spring. But because their breeding-grounds are warming up so rapidly, spring is arriving much earlier in those northerly regions. As a result, the annual glut of insects, on which knots and other Arctic-breeding waders feed their hungry chicks, are also emerging earlier and earlier each year. This means that if the knots return on the same date as they usually do, by the time their chicks have hatched, the insects have dwindled to almost nothing. The chicks starve, and year on year the population begins to fall. Eventually, if these breeding failures continue to happen, and numbers continue to fall, the species will be on a potentially irreversible journey towards extinction.

Arriving too late to breed is far from the only problem faced by the knot. On those journeys, as they head north in spring, and back south in autumn, they must stop and feed, to build up the energy resources they will need for the next leg of their trip. As they leave the warmer, tropical regions and pass through cooler, temperate areas, they also require larger and larger feeding areas, so that they can get enough energy.* This means that any rises in sea level – again a direct result of the climate crisis – will reduce the available feeding habitats and, as a consequence, reduce the birds' chances of survival.

The best-known stopover point in the world for red knots is Delaware Bay, along the eastern coast of the United States. This site is home to millions of horseshoe crabs, which traditionally lay their eggs at more or less exactly the same time as the birds pass through each spring. The horseshoe crab eggs are especially valuable for the knots because they are easy to digest – enabling

---

* One study revealed that flocks of knots overwintering in West Africa need a feeding area ranging from 2–16 square kilometres in size; yet on migration, when they stop off at the Wadden Sea in the Netherlands, the flocks require a much bigger area in order to forage effectively – as large as 800 square kilometres. Jutta Leyrer, Bernard Spaans, Mohamed Camara and Theunis Piersma, 'Small home ranges and high site fidelity in red knots (*Calidris c. canutus*) wintering on the Banc d'Arguin, Mauritania', *Journal of Ornithology* 147 (2), 2006.

the birds to double their weight during the week or two they stop to feed here.

This is no coincidence: the red knot has evolved to time its migration in order to take advantage of this glut of energy-rich food, which, like the avian equivalent of a motorway service station, provides them with the fuel they need to continue their journey.[26] As a result, between half and three-quarters of the entire '*rufa*' subspecies of red knot – the population that breeds in Arctic Canada – passes through Delaware Bay each year.[27] But changes in the local climate mean that the crabs are now spawning earlier than they used to, so again there is a danger that the knots will mistime their arrival, and find that there is not enough for them to eat.

Likewise, on the other side of the North Atlantic, Atlantic puffins – many people's favourite seabird – are struggling to find their staple diet of sand eels on which they feed their chicks, because the fish have migrated north in response to rapidly warming seas.[28]

As so often happens with the consequences of the climate crisis, these issues come on top of a whole range of other threats to birds such as the puffin and the knot, including habitat loss, pollution and the over-harvesting of natural resources – in the case of the horseshoe crab, for fertiliser and fishing bait. As a result of these threats, numbers of knots passing through Delaware Bay were beginning to decline – by as much as 75 per cent from the 1980s to the 2000s – even before changes in climate created a disconnect between the birds' arrival and the crabs laying their eggs.[29]

But why can't the knots respond to the changing climate, and adapt to suit the new conditions? Well, actually they can – and have – but unfortunately not in the right way. Because their environment, and especially the places where they breed, is getting warmer, knots are following the process known as Bergmann's Rule, in which birds living in colder environments tend to be larger, and in warmer ones, smaller, and are evolving a smaller body size and lower weight. This is directly leading to lower survival rates for the chicks. At the same time, and for the same reason, their bills are becoming progressively shorter, meaning they are unable to reach as much food as before.[30]

Actions being taken to help reverse the red knot's decline include limitations on harvesting the horseshoe crabs and their eggs, and restricting access to beaches to avoid disturbing the birds as they feed.[31] Whether these local efforts will be enough to solve a global problem is, however, in serious doubt.

Away from the poles, temperature changes as a result of the climate crisis are not – or at least not *yet* – so extreme. But even in the world's temperate regions, many species of bird are nevertheless struggling with environmental changes caused by our rapidly warming climate.

The woods and forests of North America and Europe are an ideal place for many species of songbird (members of the order *Passeriformes*, or perching birds) to breed and raise a family. During the northern hemisphere's spring and summer months, the long hours of daylight produce a veritable glut of insect food, on which a wide range of birds feed their chicks. Different species of songbird adopt two markedly diverse strategies, in order to survive long enough to pass on their genetic heritage to the next generation.

Their dilemma is best summed up by the title of a 1981 song by the English rock band the Clash: 'Should I Stay or Should I Go'.

The ones that stay are the many resident, mainly sedentary, species that breed in these northern temperate forests. In Europe, these include various thrushes (such as the Eurasian blackbird), the European robin, Eurasian wren, nuthatch, treecreeper, goldcrest and several species of tit; in North America, their counterparts include the American robin, wrens, titmice, chickadees, nuthatches and kinglets. Most of these are small and short-winged, because – apart from those that breed in the far north – they stay more or less on their home range all year round, in some cases hardly travelling any distance from where they were born and raised. Each spring, they fill their woodland homes with birdsong, starting to sing to defend their territory and win a mate as early as the New Year, with a crescendo of sound building in volume and strength as the Earth shifts towards summer in late March, at the time of the Spring Equinox. They court, mate, build a nest and lay their eggs, incubate the clutch, feed the chicks when they hatch. Then, after that first brood fledges, many often go on to have a second, and for some, a third brood, until the coming of autumn reduces the amount of available food.

At this point the birds switch their strategy from breeding to surviving the coming winter. Few live longer than two or three years; nevertheless, that is enough time, if they are fortunate, to raise more than enough young to continue their ancestral line down the generations.

Yet at the point in April when the intensity of birdsong is so high that it seems as if it could hardly get any louder or more intense, it does. New species arrive as if by magic, thronging into those same woods and forests and filling the spring air with an even greater variety of sound. Aptly, for such accomplished songsters, many go under the family name of warblers. This term was coined in the 1770s by the Welsh ornithologist Thomas Pennant, to describe a variety of small, slender-billed, insectivorous songbirds found throughout Europe and much of the Old World, traditionally grouped in the family *Sylviidae*.[32] Soon afterwards, pioneering North American ornithologists, including the Scottish-born Alexander Wilson, applied the same word to an entirely different family of superficially similar songbirds, now known as the New World Warblers, in the family *Parulidae*.[33]

Members of both these groups share many similar characteristics. They feed themselves and their young on insects; they are mostly graceful, long-winged and very active; but most of all, the vast majority of species, in both families, are long-distance migrants. Unlike their sedentary counterparts, they travel vast distances across the globe, from their northern hemisphere breeding grounds to their winter quarters, mostly in the southern hemisphere. Thus, the American warblers – together with members of other families including vireos, tanagers and tyrant flycatchers – migrate from Canada and the United States to Central and South America while their European counterparts, along with flycatchers and chats, also head south, mostly to sub-Saharan Africa.

A question that ornithologists often get asked is why would a bird, weighing perhaps 15–20 grams, risk travelling such long distances on its global journeys? Surely, the argument goes, it would do better to stay put? Yet as the Swedish ornithologist Thomas

Alerstam has pointed out, a better question would be: why don't *all* birds migrate?[34]

Migration is often the optimum strategy for a species to follow, as it gives it the best of both worlds. The birds enjoy an abundance of food, long daylight hours and less competition with other species on the northern breeding-grounds, combined with warmth and food in the southern wintering areas (remember, these global travellers actually enjoy the benefits of a second, austral summer, and never experience a true winter). They often lay smaller clutches, and have fewer broods, than resident ones, confirming that the perils of staying put for the winter generally outweigh the risks of migration.

Or – as with the life cycles of the emperor penguin and red knot – they used to. But not now. Spring is arriving earlier in these temperate latitudes, too: plants are coming into flower, and insects are emerging, as much as one month earlier than the long-term average dates.[35] Most crucially, the oak-moth caterpillars, on which European birds such as great tits and pied flycatchers feed their chicks, are also emerging much earlier than usual.

This is where the two different strategies – 'should I stay or should I go' – start to go out of balance. The resident great tits are able to adapt their life cycle, beginning to breed earlier, so that they can still find enough food for their hungry chicks when they hatch. But the pied flycatchers, which overwinter in West Africa, are unable to respond so rapidly. Because – just like the red knot – they use changes in daylight to decide when to head back north, they arrive back at roughly the same time of year: from mid-April to early May. With the shift forward of spring, by the time their chicks hatch in late May or June, the caterpillar glut is over for another year.

Unless they can rapidly adapt to migrating earlier, these long-distance travellers will continue to raise fewer young, leading to a slow but steady population decline. They might, it is true, be able to replace those caterpillars with flying insects, but with those also in free fall for other reasons, this may not be enough to save the species in the longer term.[36]

Again, for migrants in both the Old World and the New, the problems caused by the climate crisis are not happening in isolation. More than three decades ago, the US environmental scientist John Terborgh flagged up the problems facing North America's long-distance migrants in his book *Where Have All the Birds Gone?*[37] The book's cover illustration, by Audubon, is of a pair of Bachman's warblers, a species now almost certainly extinct, symbolising many other once common and familiar New World warblers which are rapidly heading in the same direction.[38]

Terborgh opens the book with a vivid account of his memories, from the early 1950s, of watching birds around his childhood home in Arlington, Virginia. As he sadly notes, for a number of reasons, including increased human population and development and a consequent loss of habitat, 'It would be impossible for a boy growing up in Arlington, Virginia, today to relive these experiences.'[39]

As he goes on to explain, those birds that are spring and summer visitors to eastern North America and spend the winter in the tropics are even more vulnerable, due to habitat loss and fragmentation on their wintering grounds. 'If these excesses continue unchecked until they run their course,' he warns, 'we shall wake up one day to a drastically altered spring – one lacking many familiar birds that we have heretofore taken for granted. If we are going to do something to prevent this, we shall have to do it soon. The year 2000 will be too late.'[40]

That last sentence feels like a punch to the stomach. Well over two decades since his deadline for action, these species are still in serious trouble. And the most shocking thing of all, is that, writing in the late 1980s, Professor Terborgh does not even mention the climate crisis as a factor in these birds' decline. That's because, at that time – little over a single human generation ago – anthropogenic climate change was not even on the agenda. Now it threatens the very existence of life on our planet.

In the cloud forests of Central America, where many of those North American warblers spend the winter, another bird is also facing up to the implications of our rapidly changing climate. But unlike the warblers, which have travelled far to be here, the resplendent quetzal is almost entirely sedentary, and can only be found between 1,200 and 2,100 metres above sea level.[41]

Arguably, the quetzal ought to have its own chapter in this book, as before the arrival of Columbus, the Mesoamerican civilisations such as the Aztecs and Mayans worshipped this bird as a god – for the Aztecs, named Quetzalcoatl. One famous legend claims that the Aztec Emperor Montezuma (see chapter 3) believed that the arrival of the Spanish conquistador Hernán Cortés signified the return of Quetzalcoatl. Consequently, it is said, the Aztecs welcomed Cortés and his men, who were then able to conquer the whole civilisation, ultimately leading to the conquest of Latin America. If that were true, then the quetzal would indeed have changed the world. But on closer examination, it appears to be a myth, made up by the Spanish in order to portray the Aztecs as primitive and naïve.[42]

Today, the quetzal retains its special status in the culture of this region, as Jonathan Evan Maslow shows in his book *Bird of Life, Bird of Death*,[43] a pioneering nature-based travelogue about Guatemala, where the quetzal is both the country's national bird and its currency. But, like the emperor penguin, red knot and the many species of migrant warbler on both sides of the Atlantic, the resplendent quetzal is under threat too. As for the penguin, its sedentary nature – and the consequent specialisation of its habitat, food and lifestyle – could soon prove to be its downfall.

Climate change does not just disrupt weather patterns at higher latitudes such as the poles: it does so at higher *altitudes* as well. As temperatures rise, so rainfall patterns change, and the delicate balance of the cloud forest, where the quetzal makes its home, begins to collapse. The future for the quetzal, and thousands of other species that thrive in the montane forest of Central America, follows the same broad pattern as the demise of the dodo, and the other oceanic island birds we saw in chapter 4. Islands function ecologically as well as geographically, and the cloud forests – surrounded by developed lowland habitats where the quetzal and others are unable to live – are in many ways no different from a physical island surrounded by sea. All over the world, there are ecological 'islands' – often mountain ranges, or simply fragments of specific habitat – that support their own highly specialised fauna and flora. If conditions change, driven by the rapid shifts brought about by the climate emergency, then the creatures that live there will not be able to survive.

One such example, the golden-cheeked warbler of Texas, is now (following the presumed extinction of Bachman's warbler, Eskimo curlew and ivory-billed woodpecker) one of the most endangered birds in the continental United States.[44] Nesting only in a particular type of juniper tree, it has suffered from the fragmentation of its habitat by development. Its cousin Kirtland's warbler, another highly endangered songbird confined to a small area of jack-pine forests in Michigan, south of the Great Lakes, also has very specific ecological requirements. The climate crisis – and its resulting changes in weather and habitat – might be the death knell for both these species.

Birds do, however, take time to actually go extinct, raising the hopes of the organisations and individuals whose aim is to save them. Then a scientific study drops a bombshell, which seems to take away all hope for the future. That's exactly what happened in March 2021 when, following a series of rampant bush fires in south-east Australia, scientists published a paper showing that male regent honeyeaters – one of Australia's rarest birds – could no longer sing their own unique song. Instead, they were reproducing the songs of other, far commoner species in the area. As a result, these males were being ignored by potential mates, hastening even further the species' imminent demise.[45]

With only about 300 regent honeyeaters remaining in the wild – spread very thinly across an area about the size of the United Kingdom – it appears that the birds have become so isolated from one another that the young males are no longer able to learn their correct song from their elders. And, as scientists, conservationists and even the notoriously climate-sceptical Australian government of the time have been forced to admit, the bush fires that had destroyed so much of the honeyeater's habitat were undoubtedly the direct result of rising temperatures caused by the global climate crisis.[46]

On a much wider scale, changes in land use as a result of climate change are likely to be extensive and unprecedented. These will affect not just rare and specialised species but common and adaptable ones as well. Large areas of land currently used for farming in the warmer areas of southern Europe are likely to go out of food production, some turning into desert or semi-desert. Farther north, huge swathes of Siberia and Canada could become suitable for growing crops or raising livestock. The consequences for the world's fauna and flora are both unknown and potentially devastating, with up to half of all species under threat of local extinction, and some disappearing altogether.[47] In some areas, the situation is far worse: in the Miombo woodlands of central and southern Africa, a predicted temperature rise of 4.5°C would result in the loss of four out of five species of mammals and over six out of every seven species of birds.[48]

Such statistics are hard for us to get our heads around. What is abundantly clear, however, is that something does need to be done – and very soon.[49] Yet, as the episode with which I opened this chapter reveals, we are more likely to engage with a problem that is both immediate and tangible – in this case, the rescue of the emperor penguins by a film crew – than something huge, longer-term and intangible. The evidence is all around us, yet we are unable – or perhaps simply unwilling – to contemplate its consequences.[50]

We are also suffering from what has been termed 'apocalypse fatigue' – the feeling that whatever we do (as individuals, societies and governments) to try to combat climate change, it simply won't be enough.[51]

Meanwhile, in the southern winter (northern summer) of 2020, a satellite survey of Antarctica brought what the scientists involved called 'good news and bad news' for the world's largest penguins.[52]

The good news was that better satellite imaging had revealed the presence of several new emperor penguin colonies. The overall number of colonies had gone up by one-fifth, while the estimated world population had risen by between 5 and 10 per cent. The bad news was that, even though there are more penguins than we had thought, they are still facing the same critical threat to their long-term existence.

The latest review of the plight of the world's birds, in early 2022, brought more bad news. It found that almost half of all bird species are known or suspected to be declining in numbers, while just 6 per cent are increasing. The phrase 'canaries in the coal mine' may be over-used, but in this case, it does seem appropriate.[53]

Cynics might dismiss the fate of the emperor penguin, the resplendent quetzal and the many other species threatened by the climate crisis as unfortunate collateral damage in the headlong drive for human economic growth. But, as we have seen throughout this book, when we mess with nature we do so at our peril: the emperors may go extinct first, but they foretell our own, not so distant, future.

We have a simple choice: will we allow the emperor penguin to follow the dodo – a bird that changed the world in the wrong way, by vanishing altogether? Or can we make the tough decisions to reverse the relentless heating produced by the climate crisis, mitigate its worst effects and ultimately save not just the emperor penguin but all life on our planet – including, of course, us?

ACKNOWLEDGEMENTS

Of all the many books I have written, this is without question the one where I have consulted and been helped by the most people, all of whom generously shared a deep knowledge and understanding of their chosen subjects with me.

So my thanks go to Tessa Boase, Esther Cheo Ying, Gordon Corera, Sarah Darwin, Frank Dikötter, Corinne Fowler, Errol Fuller, Miranda Garrett, Peter & Rosemary Grant, Tom Holland, Julian Hume, Colin Jerolmack, Carl Jones, Lesley Kinsley, Will Lawson, Tim Low, Lindsay McCrae, Tim Moreman, Jeremy Mynott, Daniel Osorio, Merle Patchett, Judith Shapiro, Michael Sheridan, Christopher Skaife and Jo Wimpenny.

I should also like to thank Charlie Gilmour, Brian Jackman, Katharine Norbury, Susie Painter, Alice Quirke and David Wright for their help and advice.

When I first came up with the idea, several birding friends, including Dominic Couzens, Mike Dilger, Ed Drewitt and Nigel Redman, helped me decide which ten birds to choose; as did my MA students at Bath Spa University, notably Rachael Bentley, Maeve Bradbury, Deborah Gray, Rachel Henson, and Debbie Rolls.

My dear friends Kevin and Donna Cox once again kindly lent me their cottage, in the heart of rural Devon, as a writing retreat, while my esteemed colleague Gail Simmons and her husband Richard Bailey read through the entire text, offering very helpful suggestions. As always, my old friend and long-time editor Graham Coster did a wonderful job of copy-editing, and huge thanks are also due to my brilliant agent Broo Doherty. Thanks, too, to Nicole Heidaripour, for her lovely chapter heading illustrations.

At Faber, I would like to thank everyone who has been involved in the commission and production of the book, including Laura Hassan (editorial), Rachael Williamson (pre-press), Connor Hutchinson and Josh Smith (publicity), John Grindrod and Phoebe Williams (marketing), Pedro Nelson (production), Anna Morrison (design), Melanie Gee (indexing), Sarah Barlow (proofreading), Dave Wright (typesetting), Louise Brice, Lizzie Bishop, Hannah Styles, Hattie Cooke (rights), and the whole of the sales team. A special thanks goes to my commissioning editor, Fred Baty, for his patience guidance and advice.

As ever, thanks to my wonderful family, here in Somerset and elsewhere: Suzanne, David & Kate (plus one!), James, Charlie, George and Daisy.

Finally, I could not have written this book without the expertise, attentiveness and sheer hard work of Lucy McRobert. As a birder as well as an historian, she was the perfect person to research the diverse stories, anecdotes, quotes, facts and figures on which I base the text.

Ten Birds That Changed the World is dedicated to Lucy and her husband Dr Rob Lambert, environmental historian at the University of Nottingham, for their friendship, support and wise counsel over the years.

NOTES

All URLs were correct when accessed at the time of writing.

INTRODUCTION

1 Eleanor Ratcliffe et al., 'Predicting the perceived restorative potential of bird sounds through acoustics and aesthetics', *Environment and Behaviour*, vol. 52, issue 4, 2020.
2 Boria Sax, *Avian Illuminations: A Cultural History of Birds* (London: Reaktion Books, 2021).
3 WWF, 'The Living Planet Report', 2018: https://www.wwf.org.uk/updates/living-planet-report-2018. See also A. Lees et al., 'State of the World's Birds. Annual Review of Environment and Resources', DOI, 2022: https://doi.org/10.1146/annurev-environ-112420-014642.
4 Worldometer: https://www.worldometers.info/world-population/ and https://www.worldometers.info/world-population/world-population-by-year/.

I RAVEN

1 As told in Bernd Heinrich's book, *Mind of the Raven* (New York: Harper Perennial, 1999, 2006).
2 Heinrich, *Mind of the Raven*.
3 See *Oxford English Dictionary* (*OED*) entry: 'raven': https://www.oed.com/view/Entry/158644?rskey=E4uNz8&result=1&isAdvanced=false#eid.
4 See Stephen Moss, *Mrs Moreau's Warbler: How Birds Got Their Names* (London: Guardian Faber, 2018).

5 Jeremy Mynott, *Birds in the Ancient World* (Oxford: Oxford University Press, 2018).

6 See 'The Ravens', Historic Royal Palaces website: https://www.hrp.org.uk/tower-of-london/whats-on/the-ravens/#gs.2c1ot4.

7 George R. R. Martin, *A Game of Thrones* (New York: Bantam Spectra, Random House, 1996). The whole sequence, known as 'A Song of Ice and Fire', tells the story of nine noble families battling it out for supreme power, in the fictional lands of Westeros and beyond. The storyline, plot and characters are based partly on the medieval Wars of the Roses and partly on the longstanding fantasy genre, with a healthy dose of sex and violence thrown in to keep the attention of a modern audience.

8 Genesis, chapter 8, verses 7–12 (King James Version, 1611).

9 'And God said, let us make man in our image, after our likeness: and let them have dominion over the fish of the sea, and over the fowl of the air, and over the cattle, and over all the earth, and over every creeping thing that creepeth upon the earth.' Genesis, Chapter 1, verse 26 (King James Version, 1611).

10 William MacGillivray, *A History of British Birds* (London: Scott, Webster and Geary, 1837–51).

11 Frank Gill, David Donsker and Pamela Rasmussen, eds, 'Family Index', IOC World Bird List Version 10.1, International Ornithologists' Union, 2020.

12 See Euring website: https://euring.org/data-and-codes/longevity-list?page=5.

13 Derek Ratcliffe, *The Raven: A Natural History in Britain and Ireland* (London: T. & A. D. Poyser, 2010).

14 Ratcliffe, *The Raven*.

15 See BirdLife International: http://datazone.birdlife.org/species/factsheet/common-raven-corvus-corax

16 Karel Voous, *Atlas of European Birds* (Amsterdam: Nelson, 1960).

17 Stanley Cramp and Christopher Perrins, eds, *Handbook of the Birds of Europe, the Middle East and North Africa. The Birds of the Western Palearctic. Volume VIII Crows to Finches* (Oxford: Oxford University Press, 1994). Incidentally, the only other species to have conquered such a wide range

of habitats is the much smaller, yet equally adaptable, Eurasian wren. See Stephen Moss, *The Wren* (London: Square Peg, 2018).

18 Ratcliffe, *The Raven*.

19 See *Danish Journal of Archaeology*, vol. 2, issue 1, 2013: https://www. tandfonline.com/doi/abs/10.1080/21662282.2013.808403?journalCode=rdja20.

20 Huginn and Muninn are inextricably linked with Odin, not just in stories and legends, but also on a wide range of archaeological artefacts including coins, helmets, brooches, fragments of tapestry, and stone carvings. See Andy Orchard, *Dictionary of Norse Myth and Legend* (London, Cassell, 1997) and Rudolf Simek, translated by Angela Hall, *Dictionary of Northern Mythology* (Woodbridge, D. S. Brewer, 2007).

21 See Nordisk Mytologi website: https://mytologi.lex.dk/Ravneguden.

22 John Lindow, *Norse Mythology: A Guide to the Gods, Heroes, Rituals, and Beliefs* (Oxford, Oxford University Press, 2001).

23 Anthony Winterbourne, *When the Norns Have Spoken: Time and Fate in Germanic Paganism* (Cranbury, New Jersey: Rosemont Publishing & Printing Corp, 2004).

24 In an endorsement of Christopher Skaife, *The Ravenmaster* (London: 4th Estate, 2018). See https://www.4thestate.co.uk/2018/11/george-rr-martin-reviews-the-ravenmaster//.

25 Heinrich, *Mind of the Raven*.

26 Heinrich, *Mind of the Raven*.

27 Heinrich, *Mind of the Raven*.

28 See Ella E. Clark, *Indian Legends of the Pacific Northwest* (Berkeley: University of California Press, 1953).

29 Early civilisations did not always make a clear distinction between crows and ravens, as noted in John M. Marzluff and Tony Angell, *In the Company of Crows and Ravens* (New Haven and London: Yale University Press, 2005).

30 Franz Boas, 'Mythology and Folk-Tales of the North American Indians', *Journal of American Folklore*, 27 (106), 1914.

31 In the Tlingit culture, of Alaska, British Columbia and the Yukon, there are two different, yet overlapping, symbols: the creator raven, and the childish

raven. See John Swanton, 'Tlingit Myths and Texts', *Bureau of American Ethnology Bulletin 39*, Smithsonian Institution, 1909.

32 Encyclopaedia of Islam: https://referenceworks.brillonline.com/entries/encyclopaedia-of-islam-3/cain-and-abel-COM_24374.

33 Mynott, *Birds in the Ancient World*. As he suggests, they may have used their sense of smell to locate the dead and dying, though that is still in dispute: see https://pubmed.ncbi.nlm.nih.gov/3960998/.

34 Edward A. Armstrong, *The Folklore of Birds* (London: Collins, 1958).

35 Quoted in Armstrong, *The Folklore of Birds*. Oddly, the numerical equivalents here are the exact reverse of the magpie verse, in which one is 'for sorrow' and two 'for joy'.

36 Revd Charles Swainson, *The Folk Lore and Provincial Names of British Birds* (London, Dialect Society, 1885). Another story suggests that a raven also helped the Vikings find – and colonise – Iceland.

37 Kristin Axelsdottir, 'The Discovery of Iceland', Viking Network, 14 August 2004.

38 Jesse Byock, *Viking Age Iceland* (London: Penguin, 2001). In the popular TV drama *Vikings*, produced by the History Channel, the character Floki the boat-builder is based on Hrafna-Flóki. See 'Iceland to play a big role in fifth season of the History channel TV series Vikings', *Iceland Magazine*, 3 March 2017.

39 Mynott, *Birds in the Ancient World*.

40 C. D. Bird, 'How the rook sees the world: a study of the social and physical cognition of *Corvus frugilegus*', PhD thesis, University of Cambridge, 2010: https://ethos.bl.uk/OrderDetails.do?uin=uk.bl.ethos.596654.

41 M. Boeckle, M. Schiestl, A. Frohnwieser, R. Gruber, R. Miller, T. Suddendorf, R. D. Gray, A. H. Taylor and N. S. Clayton, 'New Caledonian crows plan for specific future tool use', Royal Society, 2020: https://royalsocietypublishing.org/doi/10.1098/rspb.2020.1490.

42 '477+ Words to Describe Raven': https://describingwords.io/for/raven.

43 Rachel Nuwer, 'Young Ravens Rival Adult Chimps in a Big Test of General Intelligence', *Scientific American*, 2020: https://www.scientificamerican.com/article/young-ravens-rival-adult-chimps-in-a-big-test-of-general-

intelligence/. What was most surprising about this study was that by testing the ravens at four different ages – four, eight, twelve and sixteen months old – the researchers discovered that most of the tasks had been mastered at the first of these stages, with the young ravens' performances at least as good as those of adult chimpanzees and orang-utans.

44 Reported in *Science*, July 2017: 'Ravens – like humans and apes – can plan for the future': https://www.science.org/content/article/ravens-humans-and-apes-can-plan-future.

45 See Charlotte Ruhl, 'Theory of Mind', *Simply Psychology*, 7 August 2020: https://www.simplypsychology.org/theory-of-mind.html.

46 Derek Bickerton, *Adam's Tongue* (New York: Hill and Wang, 2009).

47 See Bernd Heinrich, *Ravens in Winter* (New York: Simon & Schuster, 1989, 2014).

48 For the full text of the poem, see the Poetry Foundation website: https://www.poetryfoundation.org/poems/48860/the-raven.

49 For example, 'The Ten Best Poems of All Time', *Strand Magazine* website: https://strandmag.com/the-ten-best-poems-of-all-time/.

50 William Shakespeare, *The Tragedy of Julius Caesar*, Act 5, Scene 1 (1599).

51 William Shakespeare, *The Tragedy of Macbeth*, Act 1, Scene 5 (1606).

52 William Shakespeare, *The Tragedy of Hamlet, Prince of Denmark*, Act 3, Scene 2 (1609).

53 William Shakespeare, *The Tragedy of Othello, the Moor of Venice*, Act 4, Scene 1 (1604).

54 Boria Sax, *City of Ravens* (London: Duckworth Overlook, 2011).

55 Lucinda Hawksley, 'The mysterious tale of Charles Dickens's raven', BBC Culture website, 20 August 2015: https://www.bbc.com/culture/article/20150820-the-mysterious-tale-of-charles-dickenss-raven.

56 Heinrich, *Ravens in Winter*.

57 Heinrich, *Ravens in Winter*.

58 The opening scene, which gives a good flavour of the whole episode, can be found on YouTube: https://www.youtube.com/watch?v=bLiXjaPqSyY&ab_channel=NikaGongadze.

59 J. R. R. Tolkien, *The Hobbit, or, There and Back Again* (London: George Allen and Unwin, 1937).

60 See, for example, this 'Notes and Queries' column from the *Guardian*: https://www.theguardian.com/global/2015/dec/29/weekly-notes-queries-carroll-raven-desk.

61 Heinrich, *Mind of the Raven*.

62 During the winter of 1496–7. Quoted in Ratcliffe, *The Raven*.

63 Simon Holloway, *The Historical Atlas of Breeding Birds of Britain and Ireland: 1875–1900* (Calton: T. & A. D. Poyser, 1996).

64 Abel Chapman, *The Borders and Beyond* (London: Gurney and Jackson, 1924). (Quoted in Ratcliffe, *The Raven*).

65 See Eilert Ekwall, *The Concise Oxford English Dictionary of Place Names* (Oxford: Clarendon Press, 1936).

66 Based on a bounty of 4d (roughly 1.66 new pence) per raven offered in 1731. https://www.bankofengland.co.uk/monetary-policy/inflation/inflation-calculator. See Roger Lovegrove, *Silent Fields* (Oxford, Oxford University Press, 2007).

67 William Wordsworth, *A Guide through the District of the Lakes* (Kendal: Hudson and Nicholson, 1810). Given the dangers involved in climbing a tree or steep crag to reach a raven's nest and seizing the chicks, while being harassed by the angry parents, the reward may have been deserved.

68 Lovegrove, *Silent Fields*.

69 Lovegrove, *Silent Fields*.

70 See this article by wildlife cameraman Richard Taylor-Jones, on the BBC news website, 2010: http://news.bbc.co.uk/local/kent/hi/people_and_places/nature/newsid_8727000/8727116.stm.

71 J. T. R. Sharrock (ed.), *The Atlas of Breeding Birds in Britain and Ireland* (Tring, BTO, 1976).

72 Peter Lack (ed.) *The Atlas of Wintering Birds in Britain and Ireland* (Calton: T. & A. D. Poyser, 1986).

73 D. E. Balmer et al. (eds), *Bird Atlas 2007–11: The Breeding and Wintering Birds of Britain and Ireland* (Thetford: BTO Books, 2013).

74 Balmer et al., *Bird Atlas 2007–11*.

75 Skaife, *The Ravenmaster*.

76 Felix Leigh, Thomas Crane and Ellen Houghton, *London Town* (London: Marcus Ward & Co., 1883).

77 Mentioned in Sax, *City of Ravens*.

2 PIGEON

1 See the 2005 TV documentary *War of the Birds*: https://www.imdb.com/title/tt0759946/plotsummary?ref_=tt_ov_pl. Available on YouTube: https://www.youtube.com/watch?v=sZfjbfe5SXM&ab_channel=FAMOSOXR, 10'25" into programme.

2 See also Nicholas Milton, *The Role of Birds in World War Two: How Ornithology Helped to Win the War* (Barnsley and Philadelphia: Pen and Sword, 2022).

3 PDSA Dickin Medal, PDSA Website: https://www.pdsa.org.uk/what-we-do/animal-awards-programme/pdsa-dickin-medal. This medal was created in 1943 by Maria Dickin of the UK charity PDSA (People's Dispensary for Sick Animals), to honour animals showing 'conspicuous gallantry or devotion to duty while serving or associated with any branch of the Armed Forces or Civil Defence Units'.

4 Genesis, chapter 8, verse 9 (King James Version, 1611).

5 The Noah's Ark story almost certainly has its roots in an even older tale, *The Epic of Gilgamesh*, from Ancient Sumeria, which is thought to date to at least 5,000 years ago. Interestingly, it also includes an episode when a raven and a dove (along with a swallow) are released in order to find land. See Stephanie Dalley, *Myths from Mesopotamia: Creation, The Flood, Gilgamesh, and Others (Oxford: Oxford University Press, 1989)*.

6 This convenient phrase is now considered to be a Western, colonial and Anglo-centric term; 'Levant' or 'West Asia' are the preferred terms. See 'The Middle East and the End of Empire', World History project: https://www.khanacademy.org/humanities/whp-1750/xcabef9ed3fc7da7b:unit-8-end-of-

empire-and-cold-war/xcabef9ed3fc7da7b:8-2-end-of-empire/a/the-middle-east-and-the-end-of-empire-beta.

7 See ABC News, 'Evidence Noah's Biblical Flood Happened, Says Robert Ballard', 2012. https://abcnews.go.com/Technology/evidence-suggests-biblical-great-flood-noahs-time-happened/story?id=17884533.

8 Barbara West and Ben-Xiong Zhou, 'Did chickens go north? New evidence for domestication', *Journal of Archaeological Science*, vol. 15, issue 5, September 1988: https://www.sciencedirect.com/science/article/abs/pii/0305440388900805. However, in June 2022 a new study, which re-dated bones of chickens at several sites, suggested that the domestication of the chicken happened much later than thought; perhaps 3,500 years ago. See Helena Horton, *Guardian*: https://amp.theguardian.com/science/2022/jun/06/chickens-were-first-tempted-down-from-trees-by-rice-research-suggests.

9 Leviticus, chapter 5, verse 7 (King James Version, 1611). 'And if he be not able to bring a lamb, then he shall bring for his trespass, which he hath committed, two turtledoves, or two young pigeons, unto the Lord; one for a sin offering, and the other for a burnt offering.'

10 Ruth Biasco et al., 'The earliest pigeon fanciers', Open Access report, 2014: https://www.nature.com/articles/srep05971. Overall, the remains of at least ninety different species of bird were found in and around the cave, of which the most frequent were various species of partridge, chough and swift, as well as the doves.

11 C. Vogel, *Tauben* (Berlin: Deutscher Landwirtschaftsverlag, 1992).

12 R. M. Engberg, B. Kaspers, I. Schranner, J. Kösters and U. Lösch, 'Quantification of the immunoglobulin classes IgG and IgA in the young and adult pigeon (*Columba livia*)', *Avian Pathology* 21, 1992.

13 Yotam Tepper et al., 'Signs of soil fertigation in the desert: A pigeon tower structure near Byzantine Shivta, Israel', *Journal of Arid Environments*, vol. 145, October 2017: https://www.sciencedirect.com/science/article/abs/pii/S0140196317301222?via%3Dihub.

14 Jennifer Ramsay, 'Not Just for the Birds: Pigeons in the Roman and Byzantine Near East': https://www.asor.org/anetoday/2017/11/not-just-birds.

15 Jacqueline Musset, 'Le droit de colombier en Normandie sous l'Ancien Régime', *Annales de Normandie*, vol. 34, 1984.

16 Andrew D. Blechman, *Pigeons: The Fascinating Saga of the World's Most Revered and Reviled Bird* (St Lucia: University of Queensland Press, 2007).

17 Kate Dzikiewicz, 'The Tragedy of the Most Hated Bird in America', Storage Room No. 2: Musings from the Bruce Museum Science Department, 17 April 2017: http://www.storagetwo.com/blog/2017/4/the-tragedy-of-the-most-hated-bird-in-america.

18 See Fahim Amir, 'Rats with wings', on Eurozine website, August 2013: https://www.eurozine.com/rats-with-wings/. See also Colin Jerolmack, 'How Pigeons Became Rats: The Cultural-Spatial Logic of Problem Animals', *Social Problems*, vol. 55, no. 1, 2008: https://www.jstor.org/stable/10.1525/sp.2008.55.1.72.

19 Rosemary Mosco, *A Pocket Guide to Pigeon Watching* (New York, Workman Publishing, 2021).

20 Gerard J. Holzmann and Björn Pehrson, *The Early History of Data Networks* (London: John Wiley & Sons, 1995).

21 Holzmann and Pehrson, *The Early History of Data Networks*. This clever strategy has been described as 'perhaps the earliest form of use of a negative acknowledgment signal': Gerard J. Holzmann, 'Data Communications: the first 2500 years' (New Jersey: AT&T Bell Laboratories): https://spinroot.com/gerard/pdf/hamburg94b.pdf.

22 Holzmann and Pehrson, *The Early History of Data Networks*.

23 See Stephen Moss, *The Swallow: A Biography* (London: Square Peg, 2020).

24 For more detail, see Ian Newton, *Bird Migration* (London: Collins, 2020).

25 T. Guilford, S. Roberts and D. Biro, 'Positional entropy during pigeon homing II: navigational interpretation of Bayesian latent state models', *Journal of Theoretical Biology*, 2004, https://www.robots.ox.ac.uk/~parg/pubs/bird_2.pdf .

26 Quoted in Helen Pilcher, 'Pigeons take the highway,' *Nature*, 2004: https://www.nature.com/articles/news040209-1.

27 A concept developed in the mid-twentieth century by the ethologists Niko Tinbergen and Konrad Lorenz, both of whom shared the 1973 Nobel Prize for Medicine with Karl von Frisch for their pioneering work on bird behaviour. See *Nature* website: https://www.nature.com/articles/214125920.

28 For a fuller description of the whole procedure, see Darren Naish, 'Voyeurism and Feral Pigeons', *Scientific American*, 2015: https://blogs.scientificamerican.com/tetrapod-zoology/voyeurism-and-feral-pigeons/.

29 Pliny the Elder, *Naturalis Historia* (77 AD).

30 Monica S. Cyrino, *Aphrodite: Gods and Heroes of the Ancient World* (London and New York: Routledge, 2010).

31 Luke, chapter 2, verse 24 (King James Version, 1611).

32 Matthew, chapter 3, verse 16 (King James Version, 1611).

33 See this unofficial Pablo Picasso website: https://www.pablopicasso.org/dove-of-peace.jsp.

34 As mentioned (without source) in https://www.pipa.be/en/articles/origin-belgian-racing-pigeon-rock-dove-carrier-pigeon-part-ii-9313.

35 Pliny the Elder, *Naturalis Historia*: https://www.loebclassics.com/view/pliny_elder-natural_history/1938/pb_LCL353.363.xml?readMode=recto.

36 See 'Enduring lessons from the legend of Rothschild's carrier pigeon', *Financial Times*, May 2013: https://www.ft.com/content/255b75e0-c77d-11e2-be27-00144feab7de.

37 Frederic Morton, *The Rothschilds: A Family Portrait* (London: Secker & Warburg, 1962).

38 Such as 'Carrier Pigeon Commerce, How Knowing First Helped the Rothschilds Build A Banking Empire', *Forbes*, June 2014: https://www.forbes.com/sites/samanthasharf/2014/06/18/carrier-pigeon-commerce-how-knowing-first-helped-the-rothschilds-build-a-banking-empire/?sh=7972d4f2b08f.

39 'The Pigeon Post into Paris 1870–1', University of California website: https://www.srlf.ucla.edu/exhibit/text/hist_page4.htm.

40 Andrew McNeillie, *The Magna Illustrated Guide to Pigeons of the World* (London: Magna Publishing, 1993).

41 National Museum of American History website: https://americanhistory.
si.edu/.

42 See Adam Bieniek, 'Cher Ami: the Pigeon that Saved the Lost Battalion',
2016, on the United States World War One Centennial Commission
website: https://www.worldwar1centennial.org/index.php/communicate/
press-media/wwi-centennial-news/1210-cher-ami-the-pigeon-that-saved-the-
lost-battalion.html.

43 See 'Croix de guerre', on this military history website: https://military-
history.fandom.com/wiki/Croix_de_guerre.

44 See 'Cher Ami', on the National Museum of American History collections
website: https://americanhistory.si.edu/collections/search/object/nmah_425415.

45 Bieniek, 'Cher Ami: the Pigeon that Saved the Lost Battalion'.

46 Kathleen Rooney, *Cher Ami and Major Whittlesey*: https://www.
penguinrandomhouse.com/books/624839/cher-ami-and-major-whittlesey-
by-kathleen-rooney/.

47 Wendell Levi, *The Pigeon* (Sumter, South Carolina: Levi Publishing
Company, 1977).

48 See 'Pigeons in War', Royal Pigeon Racing Association website: https://
www.rpra.org/pigeon-history/pigeons-in-war/.

49 Alexander Lee, 'Pigeon racing: A Miner's World?', *History Today*, April 2021:
https://www.historytoday.com/archive/natural-histories/pigeon-racing-
miners-world.

50 Gordon Corera, *Secret Pigeon Service: Operation Columba, Resistance and the
Struggle to Liberate Europe* (London: William Collins, 2018). Also see Jon
Day, 'Operation Columba', *London Review of Books*, 2019: https://www.lrb.
co.uk/the-paper/v41/n07/jon-day/operation-columba.

51 *War of the Birds*.

52 PDSA Dickin Medal, PDSA Website.

53 'Spy pigeon's medal fetches £9,200', BBC website, 30 November 2004:
http://news.bbc.co.uk/1/hi/uk/4054421.stm.

54 'Liberation of Europe: Pigeon Brings First Invasion News', Imperial War
Museum website: https://www.iwm.org.uk/collections/item/object/205357374.

55 See PDSA website: https://www.pdsa.org.uk/get-involved/dm75/the-relentless/duke-of-normandy.

56 Ian Herbert, 'The hero of the latest British war movie is a pigeon called Valiant. A flight of fancy? No, it's based on real life', *Independent*, 23 March 2005: https://www.independent.co.uk/news/uk/this-britain/the-hero-of-the-latest-british-war-movie-is-a-pigeon-called-valiant-a-flight-of-fancy-no-it-s-based-on-real-life-529601.html.

57 *War of the Birds.*

58 Ed.Drewitt, *Urban Peregrines* (Exeter: Pelagic Publishing, 2014).

59 National Archives website: https://discovery.nationalarchives.gov.uk/details/r/C4623103.

60 Derek Ratcliffe, *The Peregrine* (Calton: T. & A. D. Poyser, 1980).

61 National Archives website.

62 Ratcliffe, *The Peregrine.*

63 See Maneka Sanjay Gandhi, 'A Fascinating History of the Carrier Pigeons', *Kashmir Observer*, 6 July 2020: https://kashmirobserver.net/2020/07/06/a-fascinating-history-of-the-carrier-pigeons/.

64 'Pakistanis respond after "spy pigeon" detained in India', BBC News website, 2 June 2015: https://www.bbc.co.uk/news/blogs-trending-32971094.

65 'Iran arrests pigeons for spying', *Metro* website, 21 October 2008: https://metro.co.uk/2008/10/21/iran-arrests-pigeons-for-spying-56438/.

66 NBC News website: https://www.nbcnews.com/storyline/isis-uncovered/isis-executes-pigeon-bird-breeders-diyala-iraq-n287421.

67 Frank Blazich, 'In the Era of Electronic Warfare, Bring Back Pigeons', War on the Rocks website: https://warontherocks.com/2019/01/in-the-era-of-electronic-warfare-bring-back-pigeons/.

68 'The Pigeon versus the Computer: A Surprising Win for All': https://pigeonrace2009.co.za/.

69 Chris Vallance, 'Why pigeons mean peril for satellite broadband', BBC News website, 29 August 2021: https://www.bbc.co.uk/news/technology-58061230.

70 Eric Simms, *The Public Life of the Street Pigeon* (London: Hutchinson, 1979).

71 Simms, *The Public Life of the Street Pigeon*. The two stations Simms mentions, Kilburn and Finchley Road, were at the time on a branch of the Bakerloo line; this has since become part of the Jubilee line.

72 Daniel Haag-Wackernagel, 'The Feral Pigeon', University of Basel: https://stopthatpigeon.altervista.org/wp-content/uploads/2012/10/Daniel-Haag-Wackernagel-Culture-History-of-the-Pigeon-Kulturgeschichte-der-Taube.pdf.

73 'Feed the Birds', St Paul's Cathedral website: https://www.stpauls.co.uk/history-collections/history/history-highlights/feed-the-birds-1964.

74 Jen Westmoreland Bouchard, '"Feed the Birds, Tuppence a Bag . . .": A Visit to London's St Paul's Cathedral', Europe up Close website, January 2020: https://europeupclose.com/article/feed-the-birds-tuppence-a-bag-a-visit-to-londons-st-pauls-cathedral/.

75 Andrew Hosken, *Ken: The Ups and Downs of Ken Livingstone* (London: Arcadia Books, 2008).

76 John Vidal, 'London pigeon war's costly bottom line', *Guardian* website, October 2004: https://www.theguardian.com/uk/2004/oct/07/london.london.

77 Valentine Low, 'Now you risk £500 fine for feeding pigeons anywhere in Trafalgar Square', *Evening Standard* website, 10 September 2017: https://www.standard.co.uk/hp/front/now-you-risk-ps500-fine-for-feeding-pigeons-anywhere-in-trafalgar-sq-6633916.html.

78 *New York Times* website: https://www.nytimes.com/2008/05/08/world/europe/08iht-pigeon.4.12710015.html and *Observer* website: https://observer.com/2019/02/nyc-parks-ban-feeding-animals/.

79 Walter Weber, 'Pigeon Associated People Diseases', University of Nebraska, 1979: https://digitalcommons.unl.edu/cgi/viewcontent.cgi?article=1020&context=icwdmbirdcontrol.

80 Colin Jerolmack, *The Global Pigeon (Fieldwork Encounters and Discoveries)* (Chicago: University of Chicago Press, 2013).

81 Colin Jerolmack, 'How Pigeons Became Rats: The Cultural-Spatial Logic of Problem Animals', 2008: https://www.jstor.org/stable/10.1525/sp.2008.55.1.72?seq=1%2525252523metadata_info_tab_contents.

82 BBC News website, 22 January 2019: https://www.bbc.co.uk/news/uk-scotland-glasgow-west-46953707.

83 Songfacts website: https://www.songfacts.com/facts/tom-lehrer/poisoning-pigeons-in-the-park. For a recording of a live performance: https://www.youtube.com/watch?v=QNA9rQcMq00&ab_channel=TheTomLehrerWisdomChannel.

84 Melanie Rehak, 'Who Made That Twitter Bird?', *New York Times Magazine*, 8 August 2014: https://www.nytimes.com/2014/08/10/magazine/who-made-that-twitter-bird.html.

85 Doug Bowman, quoted in Rob Alderson, 'A look at the new Twitter logo and what people are reading into it', 7 June 2012: https://www.itsnicethat.com/articles/new-twitter-logo.

3 TURKEY

1 Ambrose Bierce, *The Cynic's Word Book (aka The Devil's Dictionary)* (New York: Doubleday, Page & Co., 1906).

2 See Rebecca Fraser, *The Mayflower* (New York: St Martin's Press, 2017) and Nathaniel Philbrick, *Mayflower: A Story of Courage, Community and War* (London: Penguin, 2006).

3 Verlyn Klinkenborg, 'Why Was Life So Hard for the Pilgrims?', *American History* 46 (5), December 2011.

4 John Brown, *The Pilgrim Fathers of New England and their Puritan Successors* (Pasadena, Texas: Pilgrim Publications, 1895, reprinted 1970).

5 See History website: https://www.history.com/topics/thanksgiving/history-of-thanksgiving.

6 Edward Winslow, letter of 11 December 1621, reproduced on Caleb Johnson's Mayflower History website: http://mayflowerhistory.com/letter-winslow-1621.

7 Andrew F. Smith, *The Turkey: An American Story* (Chicago and Urbana: University of Illinois Press, 2006). This is a superb account of the multi-layered history of wild and domestic turkeys, packed with information, analysis and incredible facts.

8 Albert Hazen Wright, 'Early Records of the Wild Turkey, I', *The Auk*, vol. 31, no. 3, July 1914.

9 See 'The Invention of Turkey Day', in Smith, *The Turkey*.

10 James Robertson, *American Myth, American Reality* (New York: Hill and Wang, 1980). Quoted in Smith, *The Turkey*.

11 See 'Traditional Christmas Dinners in America', Morton Williams website: https://www.mortonwilliams.com/post/traditional-christmas-dinners-in-america.

12 Nora Ephron, *Huffington Post*, November 2010 https://www.facebook.com/NoraEphron/photos/a.241541749203751/516425171715406/?type=3

13 Smith, *The Turkey*.

14 See Elizabeth Pennisi, 'Quail-like creatures were the only birds to survive the dinosaur-killing asteroid impact', *Science*, 24 May 2018: https://www.science.org/content/article/quaillike-creature-was-only-bird-survive-dinosaur-killing-asteroid-impact.

15 A. W. Schorger, *The Wild Turkey; its History and Domestication* (University of Oklahoma Press, 1966).

16 Smith, *The Turkey*. Turkeys can fly at speeds of 88 kmh per hour (55 mph).

17 'All About Birds: Wild Turkey', Cornell Lab: https://www.allaboutbirds.org/guide/Wild_Turkey/overview.

18 See Hugh A. Robertson and Barrie D. Heather, *The Hand Guide to the Birds of New Zealand* (Penguin Random House New Zealand, 1999, 2015).

19 Ralph Thomson, 'Richmond Park and the Georgian access controversy', National Archives, 23 June 2021: https://blog.nationalarchives.gov.uk/richmond-park-and-the-georgian-access-controversy/.

20 See '4 Facts about Declining Turkey Populations', NWTF website: https://www.nwtf.org/content-hub/4-facts-about-declining-turkey-populations.

21 See M. Shahbandeh, 'Number of turkeys worldwide from 1990 to 2020', Statista website, 24 January 2022: https://www.statista.com/statistics/1108972/number-of-turkeys-worldwide/.

22 However, in June 2022 a new study which re-dated bones of chickens at several sites suggested that the domestication of the chicken happened

much later than thought: perhaps 3,500 years ago. See Helena Horton, *Guardian*, 6 June 2022: https://amp.theguardian.com/science/2022/jun/06/chickens-were-first-tempted-down-from-trees-by-rice-research-suggests.

23 Michael Price, 'The turkey on your Thanksgiving table is older than you think', *Science*, November 2021, 2018: https://www.science.org/content/article/turkey-your-thanksgiving-table-older-you-think. The Muscovy duck *Cairina moschata* is the only other American bird to have been domesticated on any scale, but it has never become common or widespread.

24 Smith, *The Turkey*.

25 Linda S. Cordell, *Ancient Pueblo Peoples* (Washington DC: St Remy Press and Smithsonian Institution, 1994). So called because it lies at the intersection of four different states: Colorado, Utah, Arizona and New Mexico. Note that the question of exactly when and how turkeys was domesticated is muddied by the lack of written evidence and the reliance on second-hand sources, mostly written many centuries later by the Spanish conquerors.

26 Erin Kennedy Thornton et al., 'Earliest Mexican Turkeys (*Meleagris gallopavo*) in the Maya Region: Implications for Pre-Hispanic Animal Trade and the Timing of Turkey Domestication', 2012: https://journals.plos.org/plosone/article?id=10.1371/journal.pone.0042630.

27 Thornton et al., 'Earliest Mexican Turkeys'.

28 David Malakoff, 'We used to revere turkeys, not eat them', *Science* website, 25 November 2015: https://www.science.org/content/article/we-used-revere-turkeys-not-eat-them. This is just one of many examples of similar turkey burial sites throughout the southern USA.

29 Schorger, *The Wild Turkey*.

30 Thornton et al., 'Earliest Mexican Turkeys'.

31 Dr R. Kyle Bocinsky, Washington State University; quoted in Malakoff, 'We used to revere turkeys, not eat them'.

32 Schorger, *The Wild Turkey*.

33 Schorger, *The Wild Turkey*.

34 Schorger, *The Wild Turkey*.

35 See M. F. Fuller and N. J. Benevenga (eds), *The Encyclopaedia of Farm Animal Nutrition* (Wallingford: CABI, 2004). Strickland did travel to the Americas as a young man, and might well have brought back some of the first birds, though it is likely that he was just one of a number of people importing them. However, the entry on William Strickland in the *Dictionary of National Biography* does not mention the turkey.

36 David Gentilcore, *Food and Health in Early Modern Europe: Diet, Medicine and Society, 1450–1800* (London: Bloomsbury Academic, 2015). Quoted in Heather Horn, 'How Turkey Went Global', *The Atlantic*, 26 November 2015: https://www.theatlantic.com/international/archive/2015/11/turkey-history-world-thanksgiving/417849/.

37 'Where Did the Domestic Turkey Come From?', All About Birds: Wild Turkey, the Cornell Lab, https://www.allaboutbirds.org/news/where-did-the-domestic-turkey-come-from/.

38 Smith, *The Turkey*.

39 Daniel Defoe, *A Tour Thro' the Whole Island of Great Britain* (1724–7).

40 Smith, *The Turkey*.

41 Smith, *The Turkey*.

42 John Gay, *Fables* (London: J. F. and C. Rivington, 1792). Quoted in Smith, *The Turkey*.

43 See 'Charles Dickens and the birth of the classic English Christmas dinner', on *The Conversation* website: https://theconversation.com/charles-dickens-and-the-birth-of-the-classic-english-christmas-dinner-108116.

44 See Robert Krulwich, 'Why a Turkey is Called a Turkey', NPR website, 27, November 2008: https://www.npr.org/templates/story/story.php?storyId=97541602&t=1644228448947.

45 Schorger, *The Wild Turkey*.

46 James A. Jobling, *The Helm Dictionary of Scientific Bird Names* (London: Christopher Helm, 2010).

47 See *Oxford English Dictionary*, 'Turkey': https://www.oed.com/view/Entry/207632?rskey=5UbJY6&result=2&isAdvanced=false#eid.

48 William Strachey, *History of the Travaile into Virginia Britannica* (London: Hakluyt Society, 1849, written in 1612).

49 Albert Hazen Wright, 'Early Records of the Wild Turkey, II', *The Auk*, vol. 31, no. 4, October 1914.

50 Schorger, *The Wild Turkey*.

51 Schorger, *The Wild Turkey*.

52 Thomas Hamilton, *Men and Manners in America* (Philadelphia: Augustus M. Kelley, 1833). Quoted in Schorger, *The Wild Turkey*.

53 Wright, 'Early Records of the Wild Turkey, II'.

54 Mark Cocker and David Tipling, *Birds and People* (London: Jonathan Cape, 2013).

55 *Oxford English Dictionary*.

56 Interestingly, the phrase also has racist connotations, as this article explains: Merrill Perlman, 'Let's not "talk turkey"', *Columbia Journalism Review*, 23 November 2015: https://www.cjr.org/language_corner/lets_not_talk_turkey.php.

57 Anyone who has ever listened to John Lennon's harrowing song 'Cold Turkey' will be left in no doubt as to the agonies endured during this process. See this clip on YouTube: https://www.youtube.com/watch?v=2C6ThAaxrWw&ab_channel=johnlennon.

58 Quoted in Schorger, *The Wild Turkey*.

59 Wright, 'Early Records of the Wild Turkey, II'.

60 Ryan Johnson, 'Global turkey meat market: Key findings and insights', the Poultry Site, 19 May 2018: https://www.thepoultrysite.com/news/2018/05/global-turkey-meat-market-key-findings-and-insights.

61 'Turkey by the numbers', National Turkey Federation website: https://www.eatturkey.org/turkeystats/.

62 The Poultry Site, 2018.

63 The Poultry Site, 2018.

64 Lorraine Murray, 'Consider the turkey', Advocates for Animals, 2007, accessed via Saving Earth (Encyclopaedia Britannica) website: https://www.britannica.com/explore/savingearth/consider-the-turkey-3.

65 *Lancaster Farming*, Pennsylvania; quoted in Murray, 2007. Ibid.

66 Murray, 'Consider the turkey'.

67 See 'The Night I Shaved the Turkey and other Thanksgiving Disasters', New England Today website: https://newengland.com/today/living/new-england-nostalgia/thanksgiving-disasters/.

68 See the UK Government's Food Standards Agency website: https://www.food.gov.uk/news-alerts/news/avoid-the-unwanted-gift-of-food-poisoning-this-christmas.

69 See 'Turkey Trouble: Home cooks risk food poisoning from washing their Christmas bird', University of Manchester website, 22 December 2014: https://www.manchester.ac.uk/discover/news/turkey-trouble-home-cooks-risk-food-poisoning-from-washing-their-christmas-bird/.

70 Helen Fielding, *Bridget Jones's Diary* (London: Picador, 1996).

71 See BBC News website: https://www.bbc.co.uk/news/uk-england-london-20908427. This eventually resulted in the conviction and jailing of the chef and manager, who had not only served contaminated turkey meat but had then also fabricated food safety records: https://www.bbc.co.uk/news/uk-england-london-30954210.

72 'Most Common Sources Of Food Poisoning From Thanksgiving Dinner', Wallace Law website: https://www.bawallacelaw.com/most-common-sources-of-food-poisoning-from-thanksgiving-dinner/.

73 See 'Meat Eater's Guide to Climate Change and Health': https://www.ewg.org/meateatersguide/eat-smart/.

74 See Live Kindly website: https://www.livekindly.co/turkey-christmas-dinner-double-emissions-vegan-roast/.

75 See the Human League website: https://thehumaneleague.org/article/lab-grown-meat.

76 See Jan Dutkiewicz and Gabriel N. Rosenberg, 'Man v food: is lab-grown meat really going to solve our nasty agriculture problem?', *Guardian*, 29 July 2021: https://www.theguardian.com/news/2021/jul/29/lab-grown-meat-factory-farms-industrial-agriculture-animals.

77 John Josselyn, *New-England's Rarities Discovered: In Birds, Beasts, Fishes, Serpents, And Plants of That Country* (London: G. Widdowes, 1672). Quoted in Smith, *The Turkey*.

78 Zadock Thompson, *History of Vermont, Natural, Civil and* Statistical, 1842. Quoted in Albert Hazen Wright, 'Early Records of the Wild Turkey, III', *The Auk*, vol. 32, no. 1, 1915.

79 Audubon, *The Birds of America*. See 'Wild Turkey' on the Audubon website: https://www.audubon.org/birds-of-america/wild-turkey.

80 T. Edward Nickens, 'Wild Turkey on the Rocks?', *Audubon Magazine*, November–December 2013: https://www.audubon.org/magazine/wild-turkey-rocks.

81 See NWTF website: https://www.nwtf.org/content-hub/4-facts-about-declining-turkey-populations.

82 Nickens, 'Wild Turkey on the Rocks?'.

4 DODO

1 Will Cuppy, quoted in *The Dodo: The History and Legacy of the Extinct Flightless Bird* (Charles River Editors, 2020).

2 See Roisin Kiberd, 'The Dodo Didn't Look Like You Think It Does', on the Vice website: https://www.vice.com/en/article/vvbqq9/the-dodo-didnt-look-like-you-think-it-does.

3 See Julian P. Hume, 'The history of the dodo *Raphus cucullatus* and the penguin of Mauritius', *Historical Biology*, 18:2, 2006: http://julianhume.co.uk/wp-content/uploads/2010/07/History-of-the-dodo-Hume.pdf. Savery painted at least ten pictures that featured the dodo, more than any other contemporary artist.

4 See Alan Grihault, *Dodo: The Bird Behind the Legend* (Mauritius: IPC Ltd, 2005).

5 Though, according to the *Oxford English Dictionary*, this phrase is little more than a century old, having first appeared in print as recently as 1904. It derives from a similarly alliterative phrase, 'as dead as a doornail'.

6 *Oxford English Dictionary*.

7 Errol Fuller, *Dodo: From Extinction to Icon* (London: HarperCollins, 2002).

8 Fuller, *Dodo: From Extinction to Icon.*

9 See *The Ecologist* website, 22 October 2019: https://theecologist.org/2019/oct/22/age-extinction.

10 For a thoroughly detailed and comprehensive account of the various accounts of the dodo and its closest relative the solitaire, covering virtually all the primary and secondary sources, see Jolyon C. Parish, *The Dodo and the Solitaire: A Natural History* (Bloomington and Indianapolis, Indiana University Press, 2013).

11 From *A True Report of the gainefull, prosperous and speedy voyage to Java in the East Indies* (London, 1599).

12 From Jacob Corneliszoon van Neck, *Het Tweede Boeck* (Amsterdam, 1601).

13 Nehemiah Grew, *Musaeum Regalis Societatis: Or, a catalogue and description of the natural and artificial rarities belonging to the* Royal Society, *and preserved at Gresham Colledge* [*sic*] (London, 1685).

14 Richard Owen, *Observations on the Dodo* (London: Proc. Zool. Soc., 1846), pp. 51–3.

15 Johannes Theodor Reinhardt, 'Nøjere oplysning om det i Kjøbenhavn fundne Drontehoved', *Nat. Tidssk. Krøyer.* IV, 1842–3, pp. 71–2.

16 Hugh Edwin Strickland and Alexander Gordon Melville, *The Dodo and its Kindred; or the History, Affinities, and Osteology of the Dodo, Solitaire, and Other Extinct Birds of the Islands Mauritius, Rodriguez, and Bourbon* (London, Reeve, Benham, and Reeve, 1848).

17 B. Shapiro et al., 'Flight of the Dodo', *Science*, 295 (5560), 2002, p. 1683.

18 J. P. Hume, *Extinct Birds*, 2nd ed. (London, Helm, 2017). Only the rails (*Ralliformes*) and the parrots (*Psittaciformes*) have suffered comparable losses.

19 Jeremy Hance, 'Caught in the crossfire: little dodo nears extinction', *Guardian*, 9 April 2018: https://www.theguardian.com/environment/radical-conservation/2018/apr/09/little-dodo-manumea-tooth-billed-pigeon-samoa-critically-endangered-hunting.

20 Anthony S. Cheke and Julian P. Hume, *Lost Land of the Dodo: An Ecological History of Mauritius, Réunion & Rodrigues* (New Haven and London: T. & A. D. Poyser, 2008).

21 Volkert Evertsz, quoted in Anthony S. Cheke, 'The Dodo's last island', Royal Society of Arts and Sciences of Mauritius, 2004.

22 Michael Blencowe, *Gone: A Search of What Remains of the World's Extinct Creatures* (London: Leaping Hare Press, 2020).

23 Julian Hume estimates that only a handful of dodos – perhaps as few as four or five – were transported out of Mauritius. His co-author, Anthony Cheke, thinks the number may have been slightly higher, at eleven or more. Cheke and Hume, *Lost Land of the Dodo*.

24 Errol Fuller, *Extinct Birds* (Oxford: Oxford University Press, 2000).

25 Fuller, *Extinct Birds*.

26 Genesis, chapter 1, verse 21 (King James Version, 1611).

27 Arthur O. Lovejoy, *The Great Chain of Being: A Study of the History of an Idea* (New York: Harper, 1936, 1960).

28 Samuel T. Turvey and Anthony S. Cheke, 'Dead as a dodo: The fortuitous rise to fame of an extinction icon', *Historical Biology*, vol. 20, no. 2, June 2008.

29 Georges Cuvier, 'Memoir on the Species of Elephants, Both Living and Fossil', 1796, 1998: https://www.tandfonline.com/doi/abs/10.1080/02724634.1998.10011112.

30 Colin Barras, 'How humanity first killed dodo, then lost it as well', *Panorama* website, 12 April 2016: https://m.theindependentbd.com/arcprint/details/40404/2016-04-12.

31 Barras, 'How humanity first killed dodo, then lost it as well'.

32 Fuller, *Dodo: From Extinction to Icon*.

33 Oxford Museum of Natural History website: https://oumnh.ox.ac.uk/the-oxford-dodo.

34 University of Copenhagen Natural History Museum of Denmark website: https://snm.ku.dk/english/exhibitions/precious_things/. This was the same specimen that Johannes Reinhardt examined in order to deduce the dodo's relationship to pigeons and doves.

35 The National Museum of the Czech Republic website: https://www.nm.cz/
en/about-us/science-and-research/collection-of-birds.

36 Often translated into English as 'Pond of Dreams' but, according to Julian
Hume, actually referring to an edible plant, brought to Mauritius by
indentured workers from India. Cheke and Hume, *Lost Land of the Dodo*.

37 Julian P. Hume and Christine Taylor, 'A gift from Mauritius: William
Curtis, George Clark and the Dodo', *Journal of the History of Collections*,
vol. 29 no. 3, 2017: http://julianhume.co.uk/wp-content/uploads/2010/07/
Hume-Taylor-Curtis-Clark-dodo.pdf. Clark did keep a few bones back for
himself, perhaps out of sentiment, and on his death in 1873 left them to
his children. Almost fifty years later, in 1921, having fallen on hard times,
his surviving daughter Edith sold them to Thomas Parkin, the founder
and President of the Hastings and St. Leonards Natural History Society in
Sussex. (See Grihault, *Dodo: The Bird Behind the Legend*.)

38 Julian Hume, Cheke and Hume, *Lost Land of the Dodo*.

39 Lewis Carroll, *Alice's Adventures in Wonderland* (London: Macmillan,
1865). 'Alice in Wonderland', as it is usually known, has never been out of
print, has been translated into well over 100 languages, and sold millions
of copies. Yet just twenty-three copies of the original first edition have
survived, including Dodgson's own copy, which sold for $1.54 million at
auction in New York in 1998 – then a record price for a children's book:
https://www.nytimes.com/1998/12/11/nyregion/auction-record-for-an-
original-alice.html.

40 W. J. Broderip, *The Penny Magazine* (London: Society for the Diffusion of
Useful Knowledge, 1833). Later reprinted in *The Penny Cyclopaedia*, 1837.

41 Turvey and Cheke, 'Dead as a dodo'.

42 Genesis, chapter 1, verse 28 (King James Version, 1611). Quoted in
Strickland and Melville, *The Dodo and its Kindred*.

43 Errol Fuller, *The Great Auk* (Kent: Errol Fuller, 1998).

44 Strickland and Melville, *The Dodo and its Kindred*.

45 Turvey and Cheke, 'Dead as a dodo'. Ironically, both the species they
singled out as on the verge of disappearing have now been officially declared

extinct. See Ian Sample, 'Yangtze river dolphin driven to extinction', *Guardian* website, 8 August 2007: https://www.theguardian.com/ environment/2007/aug/08/endangeredspecies.conservation. Also Katharine Gammon, 'US to declare ivory-billed woodpecker and 22 more species extinct', *Guardian* website, 29 September 2021: https://www.theguardian. com/environment/2021/sep/29/us-bird-species-ivory-billed-woodpecker-extinct.

46 Cheke and Hume, *Lost Land of the Dodo*.

47 Francois Benjamin Vincent Florens, 'Conservation in Mauritius and Rodrigues: Challenges and Achievements from Two Ecologically Devastated Oceanic Islands', in *Conservation Biology: Voices from the Tropics* (London: John Wiley & Sons, 2013).

48 Heather S. Trevino, Amy L. Skibiel, Tim J. Karels and F. Stephen Dobson, 'Threats to Avifauna on Oceanic Islands'; Heather S. Trevino, Amy L. Skibiel Tim J. Karels and F. Stephen Dobson, *Conservation Biology* Vol. 21, No. 1 (Feb. 2007).

49 http://datazone.birdlife.org/sowb/casestudy/small-island-birds-are-most-at-risk-from-invasive-alien-species-.

50 R. Galbreath and D. Brown, 'The tale of the lighthouse-keeper's cat: Discovery and extinction of the Stephens Island wren (*Traversia lyalli*)', Notornis, 51(#4), 2004, pp. 193–200: https://www.birdsnz.org.nz/ publications/the-tale-of-the-lighthouse-keepers-cat-discovery-and-extinction-of-the-stephens-island-wren-traversia-lyalli/.

51 Richard P. Duncan and Tim M. Blackburn, 'Extinction and endemism in the New Zealand avifauna', *Global Ecology and Biogeography*, 2004: https:// doi.org/10.1111/j.1466-822X.2004.00132.x.

52 Morten Allentoft, quoted in Virginia Morell, 'Why Did New Zealand's Moas Go Extinct?', *Science* (American Association for the Advancement of Science, Virginia), 2014: https://www.sciencemag.org/news/2014/03/why-did-new-zealands-moas-go-extinct.

53 'Seabird recovery on Lundy', *British Birds*, vol. 112, no. 4, April 2017: https://britishbirds.co.uk/content/seabird-recovery-lundy. Also see: https://

www.theguardian.com/environment/2019/may/28/seabirds-treble-on-lundy-after-island-is-declared-rat-free.

54 E. Bell et al., 'The Isles of Scilly seabird restoration project: the eradication of brown rats (*Rattus norvegicus*) from the inhabited islands of St Agnes and Gugh, Isles of Scilly', 2019: http://www.issg.org/pdf/publications/2019_Island_Invasives/BellScilly.pdf.

55 However, the latest update, from January 2022, reveals that the first attempt did not manage to eradicate all the mice on the island. 'Gough Island restoration programme', RSPB website: https://www.rspb.org.uk/our-work/conservation/projects/gough-island-restoration-programme/.

56 'Update on Gough Island restoration', RSPB website, 13 January 2022: https://www.goughisland.com/.

57 Predator Free 2050 website: https://pf2050.co.nz/. In 2017, the then Conservation Minister Maggie Barry described this as 'the most important conservation project in the history of our country – one which will secure our native species from the threat of extinction and safeguard them for future generations'. New Zealand government press release, 25 July 2017, on Scoop website: https://www.scoop.co.nz/stories/PA1707/S00365/new-zealand-congratulated-on-predator-free-campaign.htm.

58 See Michael Greshko, *National Geographic* website, 25 July 2016: https://www.nationalgeographic.com/science/article/new-zealand-invasives-islands-rats-kiwis-conservation.

59 See Norman Myers, *The Sinking Ark* (Oxford: Pergamon Press, 1979).

60 See 'Fresh hope for one of the world's rarest raptors', Birdguides, 24 July 2021: https://www.birdguides.com/news/fresh-hope-for-one-of-worlds-rarest-raptors/.

61 Carl F. Jones et al., 'The restoration of the Mauritius Kestrel population', *Ibis*, vol. 137, 1994: https://onlinelibrary.wiley.com/doi/pdf/10.1111/j.1474-919X.1995.tb08439.x.

62 See the Durrell Wildlife Conservation Trust website: https://www.durrell.org/news/pink-pigeon-bouncing-back-from-the-brink/.

63 'Mauritian Parrot No Longer Endangered: WVI Celebrates Conservation Success', on Vet Report website: https://www.vetreport.net/2020/02/mauritian-parrot-no-longer-endangered-wvi-celebrates-conservation-success/.

64 Much of Jones's career has been spent with the Durrell Wildlife Conservation Trust, founded by the late Gerald Durrell of *My Family and Other Animals* fame and his wife Lee, in co-operation with the Mauritius government and other conservation organisations such as the Parrot Trust. See the Durrell Wildlife Conservation Trust website: https://www.durrell.org/wildlife/. For more about Jones's work, and conservation on oceanic islands, see Jamieson A. Copsey, Simon A. Black, Jim J. Groombridge and Carl G. Jones (eds.), *Species Conservation: Lessons from Islands* (Cambridge: Cambridge University Press, 2018).

65 David Quammen, *The Song of the Dodo: Island Biogeography in an Age of Extinctions* (London: Hutchinson, 1996).

66 See the Durrell Wildlife Conservation Trust website: https://www.durrell.org/news/professor-carl-jones-wins-2016-indianapolis-prize/.

67 Meg Charlton, 'What the Dodo Means to Mauritius', 2018: https://www.atlasobscura.com/articles/mauritius-and-the-dodo.

68 Jacques Germond and J. Roger Merven, *Les aventures de Maumau le dodo: souvenirs de genèse*, 1986. Quoted in Grihault, *Dodo: The Bird Behind the Legend*.

69 As shown in Fuller, *Dodo: From Extinction to Icon*.

70 The first set, exclusively featuring the dodo, was issued in 2007, and the latest, on the extinct birds of Mauritius, in 2022. Cheke and Hume, *Lost Land of the Dodo*.

71 Quammen, *The Song of the Dodo*.

72 Deborah Bird Rose, 'Double Death', 2014: https://www.multispecies-salon.org/double-death/. See also D. B. Rose, *Reports from a Wild Country: Ethics for Decolonization* (Sydney: University of NSW Press, 2004).

73 Anna Guasco, '"As dead as a dodo": Extinction narratives and multispecies justice in the museum', *Nature and Space*, August 2020: https://journals.sagepub.com/doi/full/10.1177/2514848620945310.

74 See Graham Redfearn, 'How an endangered Australian songbird is forgetting its love songs', *Guardian* website, 16 March 2021. https://www.theguardian.com/environment/2021/mar/17/how-an-endangered-australian-songbird-regent-honeyeater-is-forgetting-its-love-songs.

75 Sean Dooley, correspondence.

76 Douglas Adams and Mark Carwardine, *Last Chance to See* (London: Pan Books, 1990).

5 DARWIN'S FINCHES

1 Charles Darwin, *The Voyage of the Beagle* (London: John Murray, 1839).

2 See, for example, the official Galápagos Islands tourism website, which claims that 'Among those that struck Darwin so greatly were the finches that are now named in his honour': https://www.galapagosislands.com/info/history/charles-darwin.html.

3 See Stephen Jay Gould, *Ever Since Darwin: Reflections in Natural History* (New York: W. W. Norton, 1977).

4 For a more detailed portrait of Darwin's story, see Janet Browne, *Charles Darwin*, vols 1 and 2 (London: Pimlico, 2003).

5 Charles Darwin, *On the Origin of Species by Means of Natural Selection* (London: John Murray, 1859).

6 For more on this fascinating story, see James T. Costa, *Wallace, Darwin and the Origin of Species* (Cambridge, MA: Harvard University Press, 2014).

7 See the opening chapter of Darwin, *On the Origin of Species*.

8 Frank Gill, David Donsker and Pamela Rasmussen (eds), 'Tanagers and allies', IOC World Bird List Version 10.2, International Ornithologists' Union, July 2020.

9 *Asemospiza obscura, IUCN Red List of Threatened Species* (BirdLife International, 2016): e.T22723584A94824826. https://dx.doi.org/10.2305/IUCN.UK.2016-3.RLTS.T22723584A94824826.en. Another species, the Saint Lucia black finch (endemic to that Caribbean island), is also sometimes proposed as the ancestor of Darwin's finches, but the distances

involved make this unlikely. See Hanneke Meijer, 'Origin of the species: where did Darwin's finches come from?', *Guardian*, 30 July 2018: https://www.theguardian.com/science/2018/jul/30/origin-of-the-species-where-did-darwins-finches-come-from.

10 In recent years, some scientists have proposed that the ancestor of Darwin's finches arrived not from the east (Ecuador) but the north-east (Central America or even the Caribbean). Intriguingly, this may explain the connection with the Cocos finch. See L. F. Baptista and P. W. Trail, 'On the origin of Darwin's finches', *The Auk*, 1988. Also E. R. Funk and K. J. Burns, 'Biogeographic origins of Darwin's finches (Thraupidae: Coerebinae)', *The Auk*, 2018.

11 Sangeet Lamichhaney, 'Adaptive evolution in Darwin's Finches': https://scholar.harvard.edu/sangeet/adaptive-evolution-darwins-finches.

12 Darwin, *The Voyage of the Beagle*. The aquatic lizard is, of course, the marine iguana, endemic to the Galápagos.

13 Darwin, *The Voyage of the Beagle*.

14 Quoted in F. J. Sulloway, 'Darwin and his finches: the evolution of a legend', *Journal of the History of Biology*, 15, 1982, pp. 1–53.

15 Charles Darwin, *Journal of researches into the natural history and geology of the countries visited during the voyage of HMS* Beagle *round the world, under the Command of Capt. Fitz Roy, RN*, 2nd edition (London: John Murray, 1845). Online version: http://darwin-online.org.uk/content/frameset?itemID=F14&pageseq=1&viewtype=text.

16 Charles Darwin, Letter to Otto Zacharias, 1877. Quoted in John Van Wyhe, Darwin Online: http://darwin-online.org.uk/content/frameset?itemID=A932&viewtype=text&pageseq=1.

17 Thomas Henry Huxley, *Science and Education*, vol. 3, 1869.

18 Percy Lowe, 'The Finches of the Galápagos in Relation to Darwin's Conception of Species', talk at British Association for the Advancement of Science, Norwich, 1935. Quoted in Van Wyhe, Darwin Online.

19 Van Wyhe, Darwin Online.

20 Francis Darwin (ed.), *The Foundations of the Origin of Species: Two Essays Written in 1842 and 1844 by Charles Darwin* (Cambridge: Cambridge University Press, 1909).

21 Ted. R. Anderson, *The Life of David Lack: Father of Evolutionary Ecology* (Oxford: Oxford University Press, 2013).

22 Lack spent his spare time studying one of Britain's most familiar birds – about which he would publish what would become a bestselling book: David Lack, *The Life of the Robin* (London: H. F. & G. Witherby, 1943).

23 Anderson, *The Life of David Lack*.

24 David Lack, 'The Galápagos Finches (Geospizinae), A Study in Variation', California Academy of Sciences, San Francisco, 1945.

25 Anderson, *The Life of David Lack*.

26 David Lack, *Darwin's Finches: An Essay on the General Biological Theory of Evolution* (Cambridge: Cambridge University Press, 1947).

27 Van Wyhe, Darwin Online.

28 *The Voyage of Charles Darwin*, BBC TV series, part 6, 1978: https://www.youtube.com/watch?v=zXY-EWZU5qo&ab_channel=chiswickscience.

29 See Charles G. Sibley, 'On the phylogeny and classification of living birds', *Journal of Avian Biology*, vol. 25, no. 2, 1994: https://www.jstor.org/stable/3677024.

30 Brian Jackman, *West with the Light* (Chesham, Bradt, 2021).

31 As Thalia, now a renowned scientist, artist and author, recalls: 'For better and worse Galápagos has shaped my whole life, and every direction I have taken.' Quoted in Joel Achenbach, 'The People Who Saw Evolution', 2014: https://paw.princeton.edu/article/people-who-saw-evolution.

32 Peter Grant and Rosemary Grant, *How and Why Species Multiply: The Radiation of Darwin's Finches* (Princeton, NJ: Princeton University Press, 2008).

33 Achenbach, 'The People Who Saw Evolution'.

34 Jonathan Weiner, *The Beak of the Finch* (London: Jonathan Cape, 1994).

35 Weiner, *The Beak of the Finch*.

36 Niles Eldredge and S. J. Gould, 'Punctuated equilibria: an alternative to phyletic gradualism', in T. J. M. Schopf (ed.), *Models in Paleobiology* (San Francisco: Freeman Cooper, 1972).

37 Grant and Grant, *How and Why Species Multiply*.

38 Grant and Grant, *How and Why Species Multiply*.

39 Weiner, *The Beak of the Finch*, chapter 5.

40 Grant and Grant, *How and Why Species Multiply*. As he notes, their daughters proved surprisingly good at finding dead birds!

41 Weiner, *The Beak of the Finch*, chapter 5.

42 Peter Grant explains the crucial difference between natural selection and evolution: 'Natural selection occurs within one generation: some individuals survive or reproduce more successfully than others. Evolution occurs from one generation to the next if the selected trait such as beak size is heritable. After some delay until we had amassed the measurements, Rosemary and I showed that evolution had indeed occurred as a result of natural selection on heritably (genetically) varying traits. Much more recently we have identified some of the genes involved.' (*How and Why Species Multiply*.)

43 T. S. Schulenberg, 'The Radiations of Passerine Birds on Madagascar', in Steven M. Goodman and Jonathan P. Benstead (eds.), *The Natural History of Madagascar* (Chicago: University of Chicago Press, 2003).

44 Usually considered to be members of the finch family (*Fringillidae*), some authorities have recently placed them into their own family, *Drepanididae*. See Les Beletsky, *Birds of the World* (London: HarperCollins, 2006).

45 See 'Hawaiian honeycreepers and their evolutionary tree', blog by GrrlScientist, *Guardian*: https://www.theguardian.com/science/punctuated-equilibrium/2011/nov/02/hawaiian-honeycreepers-tangled-evolutionary-tree.

46 One of the key threats to the family is the spread of avian malaria. See Wei Liao et al., 'Mitigating Future Avian Malaria Threats to Hawaiian Forest Birds from Climate Change', *Plos One*, 6 January 2017: https://journals.plos.org/plosone/article?id=10.1371/journal.pone.0168880.

47 Alvin Powell, *The Race to Save the World's Rarest Bird: The Discovery and Death of the Po'ouli* (Mechanicsburg, PA: Stackpole Books, 2008).

48 Jon Fjeldså, Les Christidis and Per G. P. Ericson (eds), *The Largest Avian Radiation* (Barcelona: Lynx Edicions, 2020).

49 For an excellent summary of the book's findings, see Gehan de Silva Wijeyeratne, 'Book Review: *The Largest Avian Radiation*', December 2020: https://www.researchgate.net/publication/347593826_BOOK_REVIEW_ The_Largest_Avian_Radiation_The_Evolution_of_Perching_Birds_or_the_ Order_Passeriformes_Edited_by_Jon_Fjeldsa_Les_Christidis_and_Per_ GP_Ericson/citation/download.

50 Not to be confused with the American 'wood-warblers' of the family *Parulidae*, which are totally unrelated to their far less showy and colourful Old World counterparts. For an amusing take on the differences between the two groups, see here: https://www.allaboutbirds.org/news/whos-got- the-best-warblers-and-why-europe-vs-america-edition/.

51 See 'The British List' (BOU): https://bou.org.uk/wp-content/ uploads/2022/06/BOU_British_List_10th-and-54th_IOC12_1_Cat-F.pdf.

52 See Avibase: The World Bird Database: https://avibase.bsc-eoc.org/ checklist.jsp?region=WPA.

53 Fjeldså, Christidis and Ericson, *The Largest Avian Radiation*.

54 Tim Low, *Where Song Began: Australia's Birds and How They Changed the World* (New Haven: Yale University Press, 2020).

55 Sean Dooley, review of *Where Song Began*, *Sydney Morning Herald*, 23 June 2014: https://www.smh.com.au/entertainment/books/book-review-where- song-began-by-tim-low-20140623-zsj9c.html.

56 Low, *Where Song Began*.

57 Low, *Where Song Began*.

58 University of Minnesota, 'Songbirds Escaped From Australasia, Conquered Rest Of World', *ScienceDaily*, 20 July 2004: https://www.sciencedaily.com/ releases/2004/07/040720090024.htm.

59 See Galápagos Conservation Trust website: https://galapagosconservation. org.uk/wildlife/darwins-finches/. See also 'Growing parasite threat to

finches made famous by Darwin', BBC News website, 17 December 2015: https://www.bbc.co.uk/news/science-environment-35114681. It is interesting to note that the opening line of this story repeats the myth that the finches 'helped Charles Darwin refine his theory of evolution'. Some good stories never die.

6 GUANAY CORMORANT

1 Henri Weimerskirch et al., 'Foraging in Guanay cormorant and Peruvian booby, the major guano-producing seabirds in the Humboldt Current System', Marine Ecology Progress Series, 458, 2012: https://www. researchgate.net/publication/271251957_Foraging_in_Guanay_cormorant_ and_Peruvian_booby_the_major_guano-producing_seabirds_in_the_ Humboldt_Current_System.

2 C. B. Zavalaga and R. Paredes 'Foraging behaviour and diet of the guanay cormorant', *South African Journal of Marine Science*, 21:1, 1999: https:// www.tandfonline.com/doi/pdf/10.2989/025776199784125980.

3 G. T. Cushman, *Guano and the Opening of the Pacific World: A Global Ecological History* (Cambridge: Cambridge University Press, 2013).

4 Lesley J. Kinsley, 'Guano and British Victorians: an environmental history of a commodity of nature', PhD thesis, University of Bristol, 2019.: https:// research-information.bris.ac.uk/en/studentTheses/guano-and-british- victorians.

5 This verse (of which there are a number of different versions) is often said to have originated from a Victorian music-hall act, but while researching her PhD ('Guano and British Victorians'), Lesley Kinsley could find no evidence of this. She has also heard it attributed to the poet laureate Alfred, Lord Tennyson.

6 Tyntesfield, National Trust website: https://www.nationaltrust.org.uk/ tyntesfield. William Gibbs died, aged 85, in April 1875.

7 George Washington Peck, *Melbourne and the Chincha Islands: With Sketches of Lima, and a Voyage Round the World* (New York City: R. Craighead, 1854).

8 Bank of England Inflation calculator: https://www.bankofengland.co.uk/monetary-policy/inflation/inflation-calculator.

9 James Miller, *Fertile Fortune: The Story of Tyntesfield* (London: National Trust, 2003).

10 *Secrets of the National Trust*, Channel 5, December 2020: https://www.channel5.com/show/secrets-of-the-national-trust-with-alan-titchmarsh/season-4/episode-8.

11 This is part of a sometimes controversial yet essential programme launched by the National Trust, in which – in the wake of the Black Lives Matter movement, and a deeper investigation into the role of human slavery in the creation of wealth in the Western World – the organisation is finally acknowledging the more controversial aspects of its properties' history. For example, in an exhibition at Tyntesfield (https://www.nationaltrust.org.uk/tyntesfield/features/tow-high-in-transit-) and Dr Corinne Fowler's excellent project, 'Colonial Countryside: National Trust Houses Reinterpreted' (https://www.nationaltrust.org.uk/features/colonial-countryside-project).

12 *Ace Ventura: When Nature Calls*: https://www.imdb.com/title/tt0112281/.

13 Ian Fleming, *Dr No* (London: Jonathan Cape, 1958). The film starring Sean Connery, the first in the long-running James Bond franchise, appeared in 1962: https://www.imdb.com/title/tt0055928/.

14 Father Joseph de Acosta, *The natural & moral history of the Indies*, translated into English by Edward Grimston, 1604.

15 See 'Did Guano Make the Inca the World's First Conservationists?', blog by GrrlScientist: https://www.forbes.com/sites/grrlscientist/2020/08/30/did-guano-make-the-inca-the-worlds-first-conservationists/?sh=19d7c95c4060.

16 *The Myths of Mexico and Peru*: https://hackneybooks.co.uk/books/30/57/TheMythsOfMexicoAndPeru.html#ch7.

17 Cushman, *Guano and the Opening of the Pacific World*. See also G. T. Cushman, 'The Most Valuable Birds in the World: International Conservation Science and the Revival of Peru's Guano Industry, 1909–65', *Environmental History*, 10 (3), 2005, pp. 477–509.

18 Inca Garcilaso de la Vega, *Comentarios Reales de los Incas*, 1609. This may have been based on a 1553 work by the Spanish conquistador Pedro Cieza de León.

19 De la Vega, *Comentarios Reales de los Incas*.

20 Thomas Malthus, 'An Essay on the Principle of Population as It Affects the Future Improvement of Society', 1798.

21 Sir Humphry Davy, *Elements of Agricultural Chemistry* (London: Longman, 1813).

22 Megan L. Johnson, 'The English House of Gibbs in Peru's Guano Trade in the Nineteenth Century', thesis, Clemson University, 2017: https://tigerprints.clemson.edu/cgi/viewcontent.cgi?article=3798&context=all_theses. This detailed account is the source of much of the material in this section of the chapter.

23 Cushman, *Guano and the Opening of the Pacific World*. Von Liebig also popularised a theory known as 'the law of the minimum', which states that the growth of a plant crop is not dependent on the total available resource, but on the amount of the scarcest resource – what he called the 'limiting factor'. In practice, this meant that if just one essential nutrient in a fertiliser were scarce or absent, this would reduce its efficacy, and lead to much lower yields.

24 *Farmer's Magazine* (London: Rogerson and Tuxford, 1852).

25 Cushman, *Guano and the Opening of the Pacific World*.

26 Benjamin Disraeli, *Tancred; or, The New Crusade* (London: Henry Colburn, 1847).

27 A. J. Duffield, *Peru in the Guano Age; Being a short account of a recent visit to the guano deposits, with some reflections on the money they have produced and the uses to which it has been applied* (London: Richard Bentley and Son, 1877).

28 Peck, *Melbourne and the Chincha Islands*.

29 Duffield, *Peru in the Guano Age*.

30 Watt Stewart, *Chinese Bondage in Peru: A History of the Chinese Coolie in Peru, 1849–1874* (Durham: Duke University Press, 1951).

31 See David Olusoga, 'Before oil, another resource made and broke fortunes: guano', *BBC History Magazine*, no. 5, 2020.

32 Cushman, *Guano and the Opening of the Pacific World*.

33 See BirdLife International website: http://datazone.birdlife.org/species/ factsheet/guanay-cormorant-leucocarbo-bougainvilliorum. The most recent estimate of numbers (albeit from more than 20 years ago, in 1999) suggests that there are roughly 3.7 million individuals – compared with as many as 21 million in 1954: a fall of more than four-fifths in less than half a century.

34 See 'El Niño bird', on the Living Wild in South America website: http:// living-wild.net/2016/07/21/el-nino-bird/.

35 See 'LED lights reduce seabird death toll from fishing by 85 per cent, research shows', on University of Exeter website: https://www.exeter.ac.uk/ news/featurednews/title_669952_en.html.

36 Fabián M. Jaksic and José M. Fariña, 'El Niño and the Birds', Anales Instituto Patagonia (Chile), 2010: https://scielo.conicyt.cl/pdf/ainpat/v38n1/art9.pdf.

37 For a vivid eyewitness account of the modern-day guano harvest, see Neil Durfee and Ernesto Benavides, 'Holy Crap! A Trip to the World's Largest Guano-Producing Islands', on the Audubon website: https://www.audubon. org/news/holy-crap-trip-worlds-largest-guano-producing-islands. Also 'The Colony', a 2016 film installation by the acclaimed Vietnamese artist Dinh Q. Lê: https://www.ikon-gallery.org/exhibition/the-colony.

38 Courtney Sexton, 'Seabird Poop Is Worth More Than $1 Billion Annually', *Smithsonian Magazine* website, 7 August 2020: https://www. smithsonianmag.com/science-nature/seabird-poop-worth-more-1-billion- annually-180975504/. One of the more bizarre outcomes of the great 'guano rush' was the passing of the American Guano Islands Act by the US Congress in 1856 – an extraordinary piece of legislation still in force today.

39 Figures from Cushman, *Guano and the Opening of the Pacific World*.

40 See Cushman, *Guano and the Opening of the Pacific World*.

41 Cushman, *Guano and the Opening of the Pacific World*.

42 See William Furter, *A Century of Chemical Engineering* (New York: Springer, 1982).

43 Astonishingly, Cushman estimates that by 2000 the Haber-Bosch process was able to create the equivalent of all the nitrogen from guano harvested during the entire nineteenth century in just ten days. Cushman, *Guano and the Opening of the Pacific World*; quoted in Edward Posnett, *Harvest: The Hidden Histories of Seven Natural Objects* (London: Bodley Head, 2019).

44 'War Agricultural Committee, 9 September 1939', *Nature* website: https://www.nature.com/articles/144473a0.

45 Extract from the TV series *Birds Britannia*, 'Countryside Birds', BBC 4, first broadcast October 2010: https://www.bbc.co.uk/programmes/b00vssdk/episodes/guide.

46 Interviewed for *Birds Britannia*, 'Countryside Birds'.

47 For a clear and detailed account of where things went wrong, I recommend Isabella Tree's *Wilding: The return of nature to a British farm* (London: Picador, 2018).

48 Interviewed for *Birds Britannia*, 'Countryside Birds'.

49 Rachel Carson, *Silent Spring* (Boston: Houghton Mifflin Company, 1962).

50 'DDT – A Brief History and Status', US Environmental Protection Agency website: https://www.epa.gov/ingredients-used-pesticide-products/ddt-brief-history-and-status.

51 DDT and many other banned chemicals are still widely used in the developing world, where laws are far less stringent. In the past decade, Asia's vultures have plummeted in numbers by 99 per cent, thanks to the widespread use of the anti-worming drug diclofenac for cattle, which the birds then consume. See Darcy L. Ogada, Felicia Keesing and Munir Z. Virani, 'Dropping dead: causes and consequences of vulture population declines worldwide', *Annals of the New York Academy of Sciences*, 2011: https://assets.peregrinefund.org/docs/pdf/research-library/2011/2011-Ogada-vultures.pdf. Tragically, Africa's vultures may now be following suit. See Stephen Moss, 'The vultures aren't soaring over Africa – and that's bad news', *Guardian*, 13 June 2020: https://www.theguardian.com/

environment/2020/jun/13/the-vultures-arent-hovering-over-africa-and-thats-bad-news-aoe.

52 Leonard Doyle, 'America's songbirds are being wiped out by banned pesticides', *Independent*, 4 April 2008: https://www.independent.co.uk/climate-change/news/american-songbirds-are-being-wiped-out-by-banned-pesticides-804547.html.

53 Damian Carrington, 'Warning of "ecological Armageddon" after dramatic plunge in insect numbers', *Guardian*, 18 October 2017: https://www.theguardian.com/environment/2017/oct/18/warning-of-ecological-armageddon-after-dramatic-plunge-in-insect-numbers. See also Paula Kover, 'Insect "Armageddon": 5 Crucial Questions Answered', *Scientific American*, 30 October 2017: https://www.scientificamerican.com/article/insect-ldquo-armageddon-rdquo-5-crucial-questions-answered/.

54 See Dave Goulson, *Silent Earth: Averting the Insect Apocalypse* (London: Jonathan Cape, 2021).

55 Francisco Sánchez-Bayo and Kris A.G. Wyckhuys, 'Worldwide decline of the entomofauna: A review of its drivers', *Biological Conservation*, vol. 232, 2019: https://www.sciencedirect.com/science/article/abs/pii/S0006320718313636.

56 Sánchez-Bayo and Wyckhuys, 'Worldwide decline of the entomofauna'.

57 Susan S. Lang, 'Careful with that bug! It's helping deliver $57 billion a year to the US, new Cornell study reports', 1 April 2016: https://news.cornell.edu/stories/2006/04/dont-swat-those-bugs-theyre-worth-57-billion-year.

58 Justus Von Liebig, *Letters on Modern Agriculture* (London: Bradbury and Evans, 1859). Quoted in Johnson, 'The English House of Gibbs in Peru's Guano Trade in the Nineteenth Century'.

59 Cushman, *Guano and the Opening of the Pacific World*.

7 SNOWY EGRET

1 From William Wilbanks, *Forgotten Heroes: Police Officers Killed in Early Florida, 1840–1925* (Paducah, Kentucky: Turner Publishing Company, 1998).

2 Stuart B. McIver, *Death in the Everglades: The Murder of Guy Bradley, America's First Martyr to Environmentalism* (Gainesville, Florida: University Press of Florida, 2003).

3 Victoria Shearer, *It Happened in the Florida Keys* (Guilford, Connecticut: Globe Pequot Press, 2008).

4 McIver, *Death in the Everglades.*

5 'Flamingo Man Heard Him Say He'd Kill Bradley', *New York Times*, 10 June 1909.

6 McIver, *Death in the Everglades.* The various societies were, of course, named after the renowned nineteenth-century bird artist John James Audubon.

7 'Florida Fisherman Who Shot Game Warden Says It Was Done in Self-Defense', *New York Times*, 8 June 1909.

8 William Dutcher, 'Guy M. Bradley', *Bird-Lore*, vol. 7 (1905), p. 218.

9 McIver, *Death in the Everglades.*

10 BirdLife International Data Zone: http://datazone.birdlife.org/species/factsheet/snowy-egret-egretta-thula.

11 John James Audubon, *Birds of America* (1827–38).

12 'Snowy Heron, or White Egret', Audubon, *Birds of America*: https://www.audubon.org/birds-of-america/snowy-heron-or-white-egret

13 Audubon, *Birds of America.*

14 Purchasing power of one US dollar (USD) in every year from 1635 to 2020, Statista website: https://www.statista.com/statistics/1032048/value-us-dollar-since-1640/ (based on 1885 values).

15 Historical Gold Prices – 1833 to present: https://nma.org/wp-content/uploads/2016/09/historic_gold_prices_1833_pres.pdf (based on 1885 values).

16 Gilbert Pearson, quoted in William Dutcher, 'The Snowy Heron', *Bird-Lore*, vol. 6 (1905): https://en.wikisource.org/wiki/Page:Bird-lore_Vol_06.djvu/59.

NOTES

17 Quoted in Jim Huffstodt, *Everglades Lawmen: True Stories of Danger and Adventure in the Glades* (Sarasota, Florida: Pineapple Press, 2000).

18 Quoted in Jack E. Davis, *An Everglades Providence: Marjory Stoneman Douglas and the American Environmental Century* (Athens, Georgia: University of Georgia Press, 2009). As a 'collector-turned-protector', Guy Bradley was in good company: the twenty-sixth President of the United States, Theodore Roosevelt, was a former hunter who, when in office from 1901 to 1919, enacted a series of bird protection laws, and created over fifty Federal Wildlife Refuges – the very first, Pelican Island in Florida, in 1903.

19 Quoted in Tessa Boase, *Mrs Pankhurst's Purple Feather* (London: Aurum, 2018, republished in paperback 2020 as *Etta Lemon: The Woman Who Saved the Birds*).

20 Charles Cory, 1902, quoted in Mark V. Barrow, Jr, *A Passion for Birds* Princeton, New Jersey: Princeton University Press, 1998). Incidentally, Charles Cory is the man after whom Cory's shearwater is named; another potential candidate for a name change, based on the recent movement to remove politically incorrect eponyms?

21 'Cake' is actually a mistranslation – she is supposed to have said '*Qu'ils mangent de la brioche* (Let them eat brioche).' This does not, as many commentators have pointed out, make much difference. See Britannica website: https://www.britannica.com/story/did-marie-antoinette-really-say-let-them-eat-cake.

22 Kathleen Nicholson, 'Vigée Le Brun, Elisabeth-Louise' (Oxford University Press: Grove Art Online).

23 See the National Gallery of London website: https://www.nationalgallery.org.uk/paintings/international-womens-day-elisabeth-louise-vigee-le-brun.

24 Sarah Abrevaya Stein, *Plumes: Ostrich Feathers, Jews, and a Lost World of Global Commerce* (New Haven: Yale University Press, 2008).

25 Robin. W. Doughty, *Feather Fashions and Bird Preservation: A Study in Nature Protection* (Berkeley: University of California Press, 1974).

26   R. J. Moore-Colyer, 'Feathered Women and Persecuted Birds: The
     Struggle against the Plumage Trade, c. 1860–1922' (Cambridge: Cambridge
     University Press, 2000; online version, 2008: https://www.cambridge.org/
     core/journals/rural-history/article/abs/feathered-women-and-persecuted-
     birds-the-struggle-against-the-plumage-trade-c-18601922/35D6DCC1C907
     DCFF1C4AA0C36A2E3322).

27   Quoted in Moore-Colyer, 'Feathered Women and Persecuted Birds'.

28   T. H. Harrisson and P. A. D. Hollom, 'The Great Crested Grebe Enquiry',
     British Birds, vol. 26, 1932.

29   William T. Hornaday, *Our Vanishing Wild Life: Its Extermination and
     Preservation* (New York, Charles Scribner's Sons, 1913; ebook available
     http://www.gutenburg.net).

30   Malcolm Smith, *Hats: A Very Unnatural History* (Michigan State University
     Press, 2020). See also https://www.historyextra.com/period/victorian/
     victorian-hats-birds-feathered-hat-fashion/.

31   Smith, *Hats*.

32   Corey T. Callaghan, Shinichi Nakagawa and William K. Cornwell, 'Global
     abundance estimates for 9,700 bird species', Proceedings of the National
     Academy of Sciences of the United States of America, 2021: https://www.
     pnas.org/content/118/21/e2023170118.

33   'The 119th Christmas Bird Count Summary', Audubon website, 9
     December 2019: https://www.audubon.org/news/the-119th-christmas-bird-
     count-summary. The vast majority of species were from Latin America,
     where species diversity is far higher than North America. Sadly, although
     the number of counts is rising year on year, the number of individual birds
     is falling dramatically, due to continued loss of biodiversity and habitat
     across the Americas.

34   Rick Wright, 'Not Quite the Last of the Carolina Parakeet', American
     Birding Association (ABA) website, 21 February 2018: https://blog.aba.
     org/2018/02/not-quite-the-last-of-the-carolina-parakeet.html.

35   Stephen Moss, *A Bird in the Bush: A Social History of Birdwatching*
     (London: Aurum, 2004).

36 Douglas Brinkley, *The Wilderness Warrior: Theodore Roosevelt and the Crusade for America* (New York: Harper Perennial, 2009).

37 Frank Chapman, 'Birds and Bonnets', letter to the editor of *Forest and Stream* magazine, 1886. Quoted in Boase, *Mrs Pankhurst's Purple Feather*.

38 See Hansard, 26 February 1869: http://hansard.millbanksystems.com/commons/1869/feb/26/leave.

39 See James Robert Vernam Marchant, *Wild Birds Protection Acts, 1880–96* (South Carolina: BiblioLife, 2009).

40 For more on this remarkable woman, and the other founders of the RSPB, see Boase, *Mrs Pankhurst's Purple Feather*.

41 Tessa Boase, 'Five women who founded the RSPB', *Nature's Home*, RSPB, 2018: https://community.rspb.org.uk/ourwork/b/natureshomemagazine/posts/five-women-who-founded-the-rspb.

42 William Souder, 'How Two Women Ended the Deadly Feather Trade', *Smithsonian Magazine,* March 2013: https://www.smithsonianmag.com/science-nature/how-two-women-ended-the-deadly-feather-trade-23187277/.

43 See Kathy S. Mason, 'Out of Fashion: Harriet Hemenway and the Audubon Society, 1896–1905', 2002: https://www.tandfonline.com/doi/abs/10.1111/1540-6563.651014.

44 Quoted in 'Hats Off to Women Who Saved the Birds', National Public Radio, July 2015: https://www.npr.org/sections/npr-history-dept/2015/07/15/422860307/hats-off-to-women-who-saved-the-birds. Ostrich feathers were considered acceptable, as the birds were farmed and the feathers plucked, rather than being killed, for the trade.

45 Kristina Alexander, 'The Lacey Act: Protecting the Environment by Restricting Trade', Congressional Research Service Report, 14 January 2014. With fines of up to $500 for transporting prohibited items, and $200 for receiving them (equivalent to over $16,000 and $6,500 at today's values), the Lacey Act proved surprisingly effective: although illegal smuggling continued, the trade in skins and feathers began to drop rapidly.

46 Henry Ford and Samuel Crowther, *My Life and Work* (Garden City, New York: Doubleday, Page & Co., 1922).

47 See *Washington Post*, August 2020: https://www.washingtonpost.com/
climate-environment/2020/08/11/quoting-kill-mockingbird-judge-struck-
down-trumps-rollback-historic-law-protecting-birds/.

48 See the National Audubon Society website: https://www.audubon.org/
news/the-migratory-bird-treaty-act-explained.

49 H. H. Johnston, letter to *Nature*, 1 April 1920: https://www.nature.com/
articles/105168a0.

50 Karen Harris, 'The Bob: A Revolutionary and Empowering Hairstyle',
History Daily: https://historydaily.org/the-bob-a-revolutionary-and-
empowering-hairstyle.

51 Paul R. Erlich et al., 'Plume Trade', 1988: https://web.stanford.edu/group/
stanfordbirds/text/essays/Plume_Trade.html.

52 Dr Merle Patchett, 'Murderous Millinery': https://fashioningfeathers.info/
murderous-millinery/. See also Merle Patchett, 'The Biogeographies of the
Blue Bird-of-Paradise: From Sexual Selection to Sex and the City', *Journal
of Social History*, vol. 52, issue 4, 2019 (https://doi.org/10.1093/jsh/shz013
see esp. p.1079), and Merle Patchett, 'Feather-Work: A Fashioned Ostrich
Plume Embodies Hybrid and Violent Labors of Growing and Making',
GeoHumanities, 7:1, 2021, pp. 257–82: https://www.tandfonline.com/doi/fu
ll/10.1080/2373566X.2021.1904789.

53 Patchett, 'Murderous Millinery'.

54 W. H. Hudson, 'Feathered Women', 1893, SPB Leaflet no. 10. Quoted in
Philip McCouat, 'Fashion, Feathers and Animal Rights', *Journal of Art
and Society*; http://www.artinsociety.com/feathers-fashion-and-animal-
rights.html.

55 *New York Times*, 31 July 1898. Quoted in Patchett, 'Murderous Millinery'.

56 Virginia Woolf, 'The Plumage Bill', *Woman's Leader* magazine, 23 July
1920. Reprinted in *The Essays of Virginia Woolf 1919–24*, edited by Andrew
McNeillie (Boston: Mariner Books, 1991).

57 Patchett, 'Murderous Millinery'.

58 Boase, *Mrs Pankhurst's Purple Feather*.

59 Boase, *Mrs Pankhurst's Purple Feather*.

60 Quoted in Stephen Moss, *Birds Britannia* (London: HarperCollins, 2011).

61 Quoted in 'Hats Off to Women Who Saved the Birds'.

62 See RSPB Annual Report 2020: https://www.rspb.org.uk/globalassets/downloads/annual-report-2020/rspb-annual-report-08-10-2020-signedoff-interacrtive-pdf.pdf/.

63 See RSPB website: https://www.rspb.org.uk/our-work/.

64 Marjory Stoneman Douglas, *Nine Florida Stories by Marjory Stoneman Douglas* (Jacksonville: University of North Florida Press, 1990).

65 See *Wind Across the Everglades*, IMDb website: https://www.imdb.com/title/tt0052395/.

66 These include the National Fish and Wildlife Foundations Guy Bradley Award, established in 1988 to commemorate state and federal officers involved in wildlife law enforcement, and the Audubon Society's Guy Bradley Lifetime Conservation Award. See Jack E. Davis, *An Everglades Providence: Marjory Stoneman Douglas and the American Environmental Century* (Athens, Georgia: University of Georgia Press, 2009).

67 BBC News website, 13 September 2021: https://www.bbc.co.uk/news/science-environment-58508001.

68 More than 80 per cent of deaths that have occurred when defending wildlife or natural resources are in Mexico, Central and South America. Major hotspots include Brazil (with almost 450 people murdered between 2002 and 2013), Honduras and Peru. See 'Deadly Environment', a 2014 report from Global Witness: https://www.globalwitness.org/en/campaigns/environmental-activists/deadly-environment/.

69 See *Guardian*, 13 September 2021: https://www.theguardian.com/environment/2021/sep/13/colombia-12-year-old-eco-activist-refuses-to-let-death-threats-dim-passion-aoe.

70 Zane Grey, *Tales of Southern Rivers* (New York: Harper & Brothers, 1924).

1 Deena Zaru, 'The symbols of hate and far-right extremism on display in pro-Trump Capitol siege', ABC News website: https://abcnews. go.com/US/symbols-hate-extremism-display-pro-trump-capitol-siege/ story?id=75177671.

2 Bend the Arc: Jewish Action @jewishaction: https://twitter.com/ jewishaction/status/1278340461682442241?ref_src=twsrc%5Etfw%7 Ctwcamp%5Etweetembed%7Ctwterm%5E1278340461682442241%7Ctwgr%5 E%7Ctwcon%5Es1_&ref_url=https%3A%2F%2Fforward.com%2Fculture%2 F450073%2Fdid-the-trump-campaign-really-slap-a-nazi-eagle-on-a-t-shirt%2F.

3 Ruth Sarles, *A Story of America First: The Men and Women Who Opposed US Intervention in World War II* (Westport, Connecticut: Greenwood Publishing Group, 2002).

4 See Krishnadev Calamur, 'A Short History of "America First"', *The Atlantic* website: https://www.theatlantic.com/politics/archive/2017/01/trump-america-first/514037/.

5 Tim Murtaugh, Trump 2020 communications director, in an email to *USA Today*, quoted on https://eu.usatoday.com/story/news/ factcheck/2020/07/11/fact-check-trump-2020-campaign-shirt-design-similar-nazi-eagle/5414393002/.

6 The eagle was not the only Nazi-linked imagery worn on T-shirts that day: in January 2022, a year after the protests, a rioter who had worn a T-shirt with the slogan 'Camp Auschwitz' pleaded guilty to entering the Capitol.

7 Steven Heller, *The Swastika and Symbols of Hate: Extremist Iconography Today* (New York: Allworth Press, 2019). See also: 'Designing for the Far Right', *Creative Review*, 2019: https://www.creativereview.co.uk/designing-for-the-far-right/.

8 Steven Heller, quoted in Justin S. Hayes, 'Jupiter's Legacy: The Symbol of the Eagle and Thunderbolt in Antiquity and Their Appropriation by Revolutionary America and Nazi Germany', Senior Capstone Projects, 2014.

9   See the National Gallery website: https://www.nationalgallery.org.uk/
    paintings/learn-about-art/paintings-in-depth/painting-saints/recognising-
    saints-animals-and-the-body/recognising-saints-eagle.

10 'The modern coat of arms of the Russian Federation celebrates its 35th
    anniversary', the State Duma: http://duma.gov.ru/en/news/28991/.

11 See https://en.wikipedia.org/wiki/List_of_national_birds. The species
    involved include the golden, harpy, Javan hawk, steppe, African fish,
    Philippine, white-tailed and bald eagles.

12 See 'Countries With Eagles On Their Flags', World Atlas website: https://
    www.worldatlas.com/articles/countries-with-eagles-on-their-flags.html.
    Some of these are identifiable as actual species – the harpy eagle on the
    Panamanian coat-of-arms, for example – while most are generic versions of
    an unidentifiable 'eagle'. See Jack E. Davis, *The Bald Eagle: The Improbable
    Journey of America's Bird* (New York: Liveright Publishing/W. W. Norton,
    2022).

13 Janine Rogers, *Eagle* (London: Reaktion Books, 2015).

14 M. V. Stalmaster, *The Bald Eagle* (New York: Universe Books, 1987).

15 See Guinness World Records website: https://www.guinnessworldrecords.
    com/world-records/largest-birds-nest.

16 See Davis, *The Bald Eagle*.

17 William Barton, 1782, quoted in Hal Marcovitz, *Bald Eagle: The Story of
    Our National Bird* (Philadelphia: Mason Crest, 2015).

18 See 'The Great Seal of the United States', US Department of State
    Bureau of Public Affairs: https://2009-2017.state.gov/documents/
    organization/27807.pdf.

19 Benjamin Franklin, letter to Sarah Bache, 1784. Quoted on: https://
    founders.archives.gov/documents/Franklin/01-41-02-0327.

20 Franklin, letter to Sarah Bache.

21 Franklin, letter to Sarah Bache.

22 See, for example, 'Did Benjamin Franklin Want the National Bird to be
    a Turkey?', The Franklin Institute website: https://www.fi.edu/benjamin-
    franklin/franklin-national-bird.

23 The opinion of Davis in *The Bald Eagle*.

24 John James Audubon, *Journals*, 1831, Library of America.

25 Davis, *The Bald Eagle*.

26 Quoted widely, including on: https://news.google.com/
    newspapers?nid=2199&dat=19890717&id=Y5EzAAAAIBAJ&sjid=
    suYFAAAAIBAJ&pg=6526,4672472.

27 Edward J. Lenik, 'The Thunderbird Motif in Northeastern Indian Art',
    Archaeology of Eastern North America, 40, 2012, pp. 163–185: https://www.
    jstor.org/stable/23265141.

28 Isis Davis-Marks, 'Archaeologists Unearth 600-Year-Old Golden Eagle
    Sculpture at Aztec Temple', *Smithsonian Magazine*, 2 February 2021:
    https://www.smithsonianmag.com/smart-news/archaeologists-unearth-600-
    year-old-obsidian-eagle-mexico-180976894/.

29 Matthew Fielding, 'Australia's Birds and Embedded Within Aboriginal
    Culture', *DEEP (Dynamics of Eco-Evolutionary Patterns)*: https://www.
    deep-group.com/post/australia-s-birds-are-embedded-within-aboriginal-
    australian-culture.

30 2 Samuel, chapter 1, verse 23 (King James Version).

31 Revelation, chapter 4, verse 7 (King James Version). The picture is complicated
    because some Biblical references to eagles might actually be to griffon vultures,
    as in the line from Hosea warning that 'He shall come as an eagle against the
    house of the Lord, because they have transgressed my covenant, and trespassed
    against my law.' Hosea, chapter 8, verse 1 (King James Version).

32 A. B. Cook, *Zeus: A Study in Ancient Religion* (Cambridge: Cambridge
    University Press, 1914). In Homer's epics, the *Iliad* and the *Odyssey* (written
    – or at least written down – during the ninth or eighth century BC), Zeus
    had already emerged as the most important of the gods, though the eagle
    had yet to become his main companion. Soon afterwards, the two were
    inextricably linked, with the eagle becoming 'a visible manifestation, nay an
    actual embodiment of Zeus'.

33 Michael Apostoles (circa mid-fifteenth century AD), quoted in George E.
    Mylonas, 'The Eagle of Zeus', *Classical Journal*, vol. 41, no. 5, 1946.

34 The Persian Empire was centred on modern-day Iran, plus Turkey to the west, the Middle East and Egypt to the south, and parts of Pakistan and Afghanistan to the east. See 'Persian Empire', National Geographic website: https://www.nationalgeographic.org/encyclopedia/persian-empire/.

35 See 'Cyrus the Great', Britannica website: https://www.britannica.com/biography/Cyrus-the-Great.

36 *Monty Python's Life of Brian* (1979). See: https://www.youtube.com/watch?v=djZkTnJnLR0&ab_channel=SF1971.

37 Justin S. Hayes, 'Jupiter's Legacy: The Symbol of the Eagle and Thunderbolt in Antiquity and Their Appropriation by Revolutionary America and Nazi Germany', Senior Capstone Projects, 261, 2014: https://digitallibrary.vassar.edu/collections/institutional-repository/jupiters-legacy-symbol-eagle-and-thunderbolt-antiquity-and.

38 See Caesar, *Commentarii de Bello Gallico*; Dio Cassius, *Roman History*; Florus, *Epitome of Roman History*.

39 Hubert de Vries, 'Two-headed Eagle', 17 July 2011: http://www.hubert-herald.nl/TwoHeadedEagle.htm.

40 Tom Holland, personal communication.

41 See 'The Fascist Messaging of the Trump Campaign Eagle': https://hyperallergic.com/576095/facist-trump-campaign-eagle-america-first-t-shirt/.

42 Arthur Moeller van den Bruck, *Das Dritte Reich* (Berlin Ring-Verlag, 1923).

43 See Stan Lauryssens, *The Man Who Invented the Third Reich* (Stroud: The History Press, 1999).

44 See Anti-Defamation League (ADL) website: https://www.adl.org/education/references/hate-symbols/nazi-eagle.

45 Hayes, 'Jupiter's Legacy'.

46 Wagner, Richard (1995a). *Art and Politics*. Vol. 4. Lincoln (NE) and London: University of Nebraska Press. ISBN 978-0-8032-9774-6.

47 Volker Losemann, 'The Nazi Concept of Rome', in Catharine Edwards, *Roman Presences: Receptions of Rome in European Culture, 1789–1945* (Cambridge: Cambridge University Press, 1999).

48 Hayes, 'Jupiter's Legacy'.

49 Steven Morris, 'Nazi concerns denied as Barclays eagle comes down', *Guardian*, 21 Aug 2007: https://www.theguardian.com/media/2007/aug/21/advertising.business.

50 'Barclays set to drop eagle logo', Reuters website, 19 June 2007: https://www.reuters.com/article/uk-barclays-abn-eagle-idUKL1942925820070619.

51 Heller, in Hayes, 'Jupiter's Legacy'.

52 See the Boy London website: https://www.boy-london.com/collections/heritage.

53 'Fury at trendy fashion label's logo that bears an astonishing resemblance to NAZI eagle', *Daily Mail*, 5 May 2014: https://www.dailymail.co.uk/news/article-2620605/Angry-shoppers-demand-fashion-label-changes-logo-looks-like-NAZI-eagle-symbol.html.

54 Heller, in Hayes, 'Jupiter's Legacy'.

55 BBC News website, 22 April 2022: https://www.bbc.co.uk/news/world-us-canada-61192975.

56 M. B. Himle, R. G. Miltenberger, B. J. Gatheridge and C. A. Flessner, 'An evaluation of two procedures for training skills to prevent gun play in children', *Pediatrics*, 113 (1 Pt 1), January 2004: https://pubmed.ncbi.nlm.nih.gov/14702451/.

57 Paul Helmke, 'NRA's "Eddie Eagle" Doesn't Fly or Protect', *Huffington Post*, 25 May 2011: https://www.huffpost.com/entry/nras-eddie-eagle-doesnt-f_b_572285.

58 Janet Reitman, 'All-American Nazis: Inside the Rise of Fascist Youth in the US', *Rolling Stone*, 2 May 2018.

59 Maria R. Audubon (ed.), *Audubon and His Journals*, Volume 1, 1897.

60 W. Devitt Miller, from the American Museum of Natural History in New York, quoted in *Popular Science Monthly*, March 1930: https://books.google.co.uk/books?id=HCoDAAAAMBAJ&pg=PA62&redir_esc=y#v=onepage&q&f=false.

61 US Fish & Wildlife Service, Migratory Bird Treaty Act: https://www.fws.gov/birds/policies-and-regulations/laws-legislations/migratory-bird-treaty-act.php.

62 US Fish & Wildlife Service, Bald and Golden Eagle Protection Act: https://www.fws.gov/law/bald-and-golden-eagle-protection-act.

63 Carson, *Silent Spring*.

64 Davis, *The Bald Eagle*.

65 US Fish & Wildlife Service, History of Bald Eagle Decline, Protection and Recovery: https://www.fws.gov/midwest/eagle/History/index.html.

66 See 'Celebrating Our Living Symbol of Freedom', *USA Today Magazine*, vol. 144, issue 2853, June 2016.

67 'Celebrating Our Living Symbol of Freedom'.

68 Davis, *The Bald Eagle*.

69 'New Wind Energy Permits Would Raise Kill Limit of Bald Eagles But Still Boost Conservation, Officials Say', ABC News: https://abcnews.go.com/US/wind-energy-permits-raise-kill-limit-bald-eagles/story?id=38881089.

70 'Trump just gutted the law that saved American bald eagles from extinction', *Fast Company*, 12 August 2019. https://www.fastcompany.com/90389091/trump-guts-the-endangered-species-act-that-saved-bald-eagles.

71 'Mysterious death of bald eagles in US explained by bromide poisoning', *New Scientist*, 25 March 2021: https://www.newscientist.com/article/2272670-mysterious-death-of-bald-eagles-in-us-explained-by-bromide-poisoning/.

72 'Watch Donald Trump Dodge a Bald Eagle', *Time* Magazine, 9 December 2015: https://time.com/4141783/time-person-of-the-year-runner-up-donald-trump-eagle-gif/.

9 TREE SPARROW

1 Adapted from an eyewitness account in Sha Yexin, 'The Chinese Sparrows of 1958', 31 August 1997: http://www.zonaeuropa.com/20061130_1.htm. Confirmed by other eyewitness sources, including Han Suyin, 'The Sparrow Shall Fall', *New Yorker*, 10 October 1959: https://birdingbeijing.com/wp-content/uploads/2015/07/the-new-yorker-oct-10-1959.pdf.

2 Suyin, 'The Sparrow Shall Fall'.

3 See 'The Four Pests Campaign', NHD Central website:
   https://oo-08943045.nhdwebcentral.org/The_Sparrow_Massacre.

4 Suyin, 'The Sparrow Shall Fall'.

5 Esther Cheo Ying, *Black Country Girl in Red China* (London: Hutchinson,
   1980). Remarkably, over forty years after it was first published, this is still
   in print under the revised title *Black Country to Red China* (London: The
   Cresset Library, 1987), and well worth reading!

6 This, and later quotes and comments, are from a face-to-face interview with
   Esther Cheo Ying, 5 August 2021.

7 Cheo Ying, *Black Country to Red China*.

8 'RED CHINA: Death to Sparrows', *Time* Magazine, 5 May 1958.

9 'RED CHINA: Death to Sparrows'.

10 Suyin, 'The Sparrow Shall Fall'.

11 Suyin, 'The Sparrow Shall Fall'.

12 Suyin, 'The Sparrow Shall Fall'.

13 Sheldon Lou, *Sparrows, Bedbugs, and Body Shadows* (Honolulu: University
   of Hawaii Press, 2005).

14 Cheo Ying, *Black Country to Red China*.

15 Suyin, 'The Sparrow Shall Fall'.

16 Suyin, 'The Sparrow Shall Fall'.

17 See Alexander Pantsov, *Mao: The Real Story* (New York: Simon and
   Schuster, 2013).

18 Cheo Ying, *Black Country to Red China*.

19 Tso-hsin Cheng, *Distributional list of Chinese birds* (Peking: Peking Institute
   of Zoology, 1976). Though dated 1976, this was actually published in 1978.
   See Jeffery Boswall, 'Notes on the current status of ornithology in the
   People's Republic of China', *Forktail*, vol. 2, 1986.

20 For an overview of these terrible events and their causes, See *Mao's
   Great Famine*: https://www.youtube.com/watch?v=r33Q8cl87HYhttps://
   en.wikipedia.org/wiki/Great_Leap_Forward.

21 See Jonathan Mirsky, 'Unnatural Disaster', *New York Times*, 7 December 2012: https://www.nytimes.com/2012/12/09/books/review/tombstone-the-great-chinese-famine-1958-1962-by-yang-jisheng.html.

22 Jonathan Mirsky, 'China: The Shame of the Villages', *New York Review of Books*, vol. 53, no. 8, 11 May 2006.

23 Frank Dikötter, *Mao's Great Famine: The History of China's Most Devastating Catastrophe, 1958–62* (London: Bloomsbury, 2010).

24 Dikötter, *Mao's Great Famine*.

25 Quoted in Yang Jisheng, *Tombstone: The Untold Story of Mao's Great Famine* (London: Allen Lane, 2012).

26 Yang Jisheng, *Tombstone*. Despite Yang Jisheng's reputation as an historian, he was forced to publish *Tombstone* in Hong Kong. A decade after the book first appeared, it remains banned in China, though Yang suspects that tens of thousands of copies are in circulation there, having been smuggled in from abroad.

27 Yang Jisheng, *Tombstone*.

28 Tania Branigan, 'China's Great Famine: the true story', *Guardian*, 1 January 2013: https://www.theguardian.com/world/2013/jan/01/china-great-famine-book-tombstone.

29 Roderick MacFarquhar and John K. Fairbank (eds), *The Cambridge History of China, Volume 14: The People's Republic, Part 1: The Emergence of Revolutionary China, 1949–1965* (Cambridge: Cambridge University Press, 1987), p. 223.

30 That phrase is also the title of a book, *Mao's War against Nature: Politics and the Environment in Revolutionary China* by Judith Shapiro (Cambridge: Cambridge University Press, 2001).

31 Dai Qing, 2004, quoted in various sources, including: http://news.bbc.co.uk/1/hi/world/asia-pacific/3371659.stm.

32 Quoted in *Sunday Herald*, 5 July 1953: https://trove.nla.gov.au/newspaper/article/18516559.

33 *Sunday Herald*, 5 July 1953.

34 'Western Australia Makes War on Emus', British Movietone, 5 January 1933: https://www.youtube.com/watch?v=YIwAoPKeJqc.

35 Quoted in Jasper Garner Gore, 'Looking back: Australia's Emu Wars', *Australian Geographic*, 18 October 2016: https://www.australiangeographic. com.au/topics/wildlife/2016/10/on-this-day-the-emu-wars-begin/.

36 J. Denis Summers-Smith, 'Studies of West Palearctic birds 197. Tree Sparrow', *British Birds*, vol. 91, 1998: https://britishbirds.co.uk/wp-content/ uploads/article_files/V91/V91_N04/V91_N04_P124_138_A031.pdf.

37 See BTO BirdTrends, Tree Sparrow: https://www.bto.org/our-science/ publications/birdtrends/birdtrends-2018-trends-numbers-breeding-success- and-survival-uk.

38 Shuping Zhang et al., 'Habitat use of urban Tree Sparrows in the process of urbanization: Beijing as a case study', *Frontiers of Biology in China*, September 2008: https://www.researchgate.net/publication/225731390_ Habitat_use_of_urban_Tree_Sparrows_in_the_process_of_urbanization_ Beijing_as_a_case_study/figures?lo=1.

39 'Frozen sparrows the tip of the iceberg', *New Scientist*, 18 December 1993: https://www.newscientist.com/article/mg14019040-900-frozen-sparrows- the-tip-of-the-iceberg/.

40 Summers-Smith, 'Studies of West Palearctic birds 197. Tree Sparrow'.

41 'Hong Kong has higher sparrow density than the UK, survey finds', *Hong Kong Economic Journal* website 19 July 2016: https://www.ejinsight.com/eji/ article/id/1347102/20160719-survey-finds-hong-kong-has-higher-sparrow- density-than-uk. The Hong Kong Bird Watching Society, which organised the event, calculated that in urban areas there are on average 1,434 sparrows per square kilometre, mostly nesting in air vents, pipes and cavities in the walls of buildings.

42 'Urban sparrow population stable, bird watching group finds', *Standard* website, 23 August 2020: https://www.thestandard.com.hk/breaking-news/ section/4/153662/Urban-sparrow-population-stable,-bird-watching-group- finds. Because of the Covid-19 pandemic, no public volunteers were recruited in 2020, and the society had to rely on its own staff; the numbers of sparrows recorded may, as a result, have been lower.

43 J. del Hoyo, A. Elliott and D. Christie (eds), *Handbook of the Birds of the World*, vol. 14, *Bush-shrikes to Old World Sparrows* (Barcelona: Lynx Edicions, 2009).

44 See Fiona Burns et al., 'Abundance decline in the avifauna of the European Union reveals global similarities in biodiversity change', 2021: https://zenodo.org/record/5544548#.Yl52CtPMJBw. Quoted in Patrick Barkham, 'House sparrow population in Europe drops by 247 million', *Guardian*, 16 November 2021: https://www.theguardian.com/environment/2021/nov/16/house-sparrow-population-in-europe-drops-by-247m.

45 Dikötter, *Mao's Great Famine*.

46 Jung Chang and Jon Halliday, *Mao: The Untold Story* (London: Jonathan Cape, 2005).

47 Quoted in Dikötter, *Mao's Great Famine*.

48 Quoted in Dikötter, *Mao's Great Famine*.

49 Rana Mitter, review of Yang Jisheng, *Tombstone*, Guardian, 7 December 2012. https://www.theguardian.com/books/2012/dec/07/tombstone-mao-great-famine-yeng-jisheng-review.

50 Mike McCarthy, 'The sparrow that survived Mao's purge', *Independent*, 3 September 2010: https://www.independent.co.uk/climate-change/news/nature-studies-by-michael-mccarthy-the-sparrow-that-survived-mao-s-purge-2068993.html.

51 Shapiro, *Mao's War against Nature*.

52 Sha Yexin, 'The Chinese Sparrow War of 1958', EastSouthWestNorth website, 31 August 1997: http://www.zonaeuropa.com/20061130_1.htm.

53 Charlie Gilmour, personal communication.

54 Sai Kumar Sela, 'Four Pests Campaign', posted on LinkedIn: https://www.linkedin.com/pulse/four-pests-campaign-sai-kumar/?trk=read_related_article-card_title. In this short-lived revival, sparrows had been replaced by cockroaches.

55 Tim Luard, 'China follows Mao with mass cull', BBC News website, 6 January 2004: http://news.bbc.co.uk/1/hi/world/asia-pacific/3371659.stm.

10 EMPEROR PENGUIN

1 Broadcast in autumn 2018. See BBC iPlayer website: https://www.bbc. co.uk/programmes/p06mvqjc.

2 Peter T. Fretwell et al., 'An Emperor Penguin Population Estimate: The First Global, Synoptic Survey of a Species from Space', *Plos One*, 2012: https://www.ncbi.nlm.nih.gov/pmc/articles/PMC3325796/.

3 See Birdlife International website: https://www.iucnredlist.org/ species/22697752/157658053.

4 See *Aptenodytes forsteri*, IUCN Red List of Threatened Species, BirdLife International, 2020: https://www.iucnredlist.org/ species/22697752/157658053.

5 See BirdLife International website: http://datazone.birdlife.org/sowb/casestudy/ one-in-eight-of-all-bird-species-is-threatened-with-global-extinction.

6 Tony D. Williams, *The Penguins* (Oxford: Oxford University Press, 1995).

7 J. Prévost, *Ecologie du manchot empereur* (Ecology of the Emperor Penguin) (Paris: Hermann, 1961).

8 See Aaron Waters and François Blanchette, 'Modeling Huddling Penguins', National Library of Medicine, 16 November 2012: https://journals.plos. org/plosone/article?id=10.1371/journal.pone.0050277.

9 These include *Life in the Freezer* (1993), *Frozen Planet* (2011) and most recently *Dynasties* (2018). These programmes found huge audiences at the time they were broadcast, and extracts are still available on YouTube, some attracting several million views. See, for example, 'Emperor Penguins: The Greatest Wildlife Show on Earth': https://www.youtube.com/ watch?v=MfstYSUscBc&ab_channel=BBCEarth.

10 All information in this section on the breeding cycle is adapted from Williams, *The Penguins*.

11 See Ben Webster, 'Emperor penguins heading for extinction unless emissions are cut, US-Cambridge study finds', *The Times*, 4 August 2021: https://www.thetimes.co.uk/article/emperor-penguins-heading-for- extinction-unless-emissions-are-cut-us-cambridge-study-finds-8nvx3gvlh.

12 Stéphanie Jenouvrier et al., 'Projected continent-wide declines of the emperor penguin under climate change', Nature Climate Change website, 2014: https://www.nature.com/articles/nclimate2280.

13 Stéphanie Jenouvrier et al., 'The call of the emperor penguin: Legal responses to species threatened by climate change', *Global Change Biology*, 2021: https://onlinelibrary.wiley.com/doi/full/10.1111/gcb.15806.

14 Jenouvrier et al., 'The call of the emperor penguin'.

15 Edward Wilson, quoted in Bernard Stonehouse, 'The Emperor Penguin, I. Breeding Behaviour and Development', Falkland Islands Dependencies Survey, Scientific Reports No. 6., Falkland Islands Dependencies Scientific Bureau, 1953.

16 Apsley Cherry-Garrard, *The Worst Journey in the World* (London, Constable & Co., 1922).

17 Cherry-Garrard, *The Worst Journey in the World*.

18 Cherry-Garrard, *The Worst Journey in the World*.

19 Jenouvrier et al., 'The call of the emperor penguin'.

20 'Endangered and Threatened Wildlife and Plants; Threatened Species Status With Section 4(d) Rule for Emperor Penguin', US Fish and Wildlife Service, 4 August 2021, published on the US government's Federal Register: https://www.federalregister.gov/documents/2021/08/04/2021-15949/endangered-and-threatened-wildlife-and-plants-threatened-species-status-with-section-4d-rule-for.

21 Paul Voosen, 'The Arctic is warming four times faster than the rest of the world', *Science*, 14 December 2021: https://www.science.org/content/article/arctic-warming-four-times-faster-rest-world.

22 Sandy Bauers, 'Globe-spanning bird, B95 is back for another year', *Philadelphia Inquirer*, 28 May 2014.

23 See BirdLife International website: http://datazone.birdlife.org/species/factsheet/red-knot-calidris-canutus/text.

24 See Elly Pepper, 'Red Knot Listed as Threatened under the Endangered Species Act', National Resources Defense Council website, 9 December

2014: https://www.nrdc.org/experts/elly-pepper/red-knot-listed-threatened-under-endangered-species-act.

25 See '"This Is the Climate Emergency": Dozens of Sudden Deaths Reported as Canada Heat Hits Record 121°F': https://www.commondreams.org/news/2021/06/30/climate-emergency-dozens-sudden-deaths-reported-canada-heat-hits-record-121degf.

26 Red Knot *Calidris Canutus Rufa*, US Fish and Wildlife Service, August 2005: https://www.fws.gov/species/rufa-red-knot-calidris-canutus-rufa.

27 See USFWS Northeast Region Division of External Affairs, Northeast Region, US Fish and Wildlife Service: https://www.fws.gov/species/red-knot-calidris-canutus-rufa.

28 Atlantic Puffin *Fratercula arctica*: see Annette L. Fayet et al., 'Local prey shortages drive foraging costs and breeding success in a declining seabird, the Atlantic puffin', *Journal of Animal Ecology*, 21 March 2021: https://besjournals.onlinelibrary.wiley.com/doi/10.1111/1365-2656.13442.

29 'The Horseshoe Crab *Limulus Polyphemus* – A Living Fossil', US Fish & Wildlife Service, August 2006: www.fws.gov/northeast/pdf/horseshoe.fs.pdf.

30 Jan A. van Gils et al., 'Body Shrinkage Due to Arctic Warming Reduces Red Knot Fitness in Tropical Wintering Range', *Science*, American Association for the Advancement of Science, 13 May 2016: https://www.science.org/doi/10.1126/science.aad6351.

31 See 'Species Profile for Red Knot (*Calidris canutus ssp. rufa*)': https://web.archive.org/web/20111019183433/http://ecos.fws.gov/speciesProfile/profile/speciesProfile.action?spcode=B0DM.

32 Thomas Pennant, *Genera of Birds* (Edinburgh, 1773).

33 Alexander Wilson, *American Ornithology, or, The natural history of the birds of the United States* (Edinburgh: Constable and Co., 1831).

34 Thomas Alerstam, *Bird Migration* (Cambridge: Cambridge University Press, 1991).

35 Ulf Büntgen et al., 'Plants in the UK flower a month earlier under recent warming', *Proceedings of the Royal Society*, 2 February 2022: https://

royalsocietypublishing.org/doi/10.1098/rspb.2021.2456. Also see 'Butterflies emerging earlier due to rising temperatures', Natural History Museum website: https://www.nhm.ac.uk/discover/news/2016/december/butterflies-emerging-earlier-due-to-rising-temperatures.html.

36   See M. D. Burgess, et al., 'Tritrophic phenological match–mismatch in space and time', BTO Publications, April 2018: https://www.bto.org/our-science/publications/peer-reviewed-papers/tritrophic-phenological-match-mismatch-space-and-time.

37   John Terborgh, *Where Have All the Birds Gone? Essays on the Biology and Conservation of Birds That Migrate to the American Tropics* (Princeton, New Jersey: Princeton University Press, 1989).

38   See BirdLife International website: http://datazone.birdlife.org/species/factsheet/22721607.

39   Terborgh, *Where Have All the Birds Gone?*

40   Terborgh, *Where Have All the Birds Gone?*

41   In January 2022, fifty years or more since I first became aware of the existence of the quetzal, I travelled to Costa Rica to see what has been described – rightly, in my opinion – as the most beautiful bird in the world. For my account of the sighting, in the BBC Radio 4 programme *From Our Own Correspondent*, listen here (from roughly 17'30" into broadcast): https://www.bbc.co.uk/sounds/play/p0bpndd5.

42   See Susan D. Gillespie, *The Aztec Kings: The Construction of Rulership in Mexica History* (Tucson, Arizona: University of Arizona Press, 1989).

43   Jonathan Evan Maslow, *Bird of Life, Bird of Death* (New York: Simon & Schuster, 1986).

44   See 'Golden-Cheeked Warbler' (*Dendroica chrysoparia*), US Fish & Wildlife Service website: https://web.archive.org/web/20111015042905/http://ecos.fws.gov/speciesProfile/profile/speciesProfile.action?spcode=B07W.

45   Ross Crates, et al., 'Loss of vocal culture and fitness costs in a critically endangered songbird', Royal Society Publishing: https://royalsocietypublishing.org/doi/10.1098/rspb.2021.0225.

46 Report by Pallab Ghosh, 'Climate change boosted Australia bushfire risk by at least 30 per cent', BBC News website, 4 March 2020: https://www.bbc.co.uk/news/science-environment-51742646.

47 'Wildlife in a Warming World: The effects of climate change on biodiversity in WWF's Priority Places', Report by the World Wide Fund for Nature, March 2018: https://www.wwf.org.uk/sites/default/files/2018-03/WWF_Wildlife_in_a_Warming_World.pdf.

48 WWF, 'Wildlife in a Warming World'.

49 Just as I have not examined the specific causes of the current climate crisis, neither am I going to propose solutions; these are, however, being considered in great depth. Whether they will be implemented in time to save the planet – or at least those species, including us, that depend on it – is up for debate. See '10 Solutions for Climate Change', *Scientific American*: https://www.scientificamerican.com/article/10-solutions-for-climate-change/.

50 This phenomenon was viciously satirised in the Oscar-nominated 2021 Netflix movie *Don't Look Up*, whose story featured a comet threatening to destroy planet Earth, and the 'business-as-usual' attitude of governments and big corporations: https://www.imdb.com/title/tt11286314/.

51 See, for example: Damaris Zehner, 'Apocalypse Fatigue, Selective Inattention, and Fatalism: The Psychology of Climate Change', Resilience website, 27 January 2020: https://www.resilience.org/stories/2020-01-27/apocalypse-fatigue-selective-inattention-and-fatalism-the-psychology-of-climate-change/.

52 Peter T. Fretwell and Philip N. Trathan, 'Discovery of new colonies by *Sentinel2* reveals good and bad news for emperor penguins', *Remote Sensing in Ecology and Conservation*, vol. 7, issue 2, 4 August 2020: https://zslpublications.onlinelibrary.wiley.com/doi/10.1002/rse2.176.

53 Alexander C. Lees et al., 'State of the World's Birds', *Annual Review of Environment and Resources*, vol. 472022: https://www.annualreviews.org/doi/10.1146/annurev-environ-112420-014642. Also reported by Damian

Carrington, *Guardian*, 5 May 2022: https://www.theguardian.com/
environment/2022/may/05/canaries-in-the-coalmine-loss-of-birds-signals-
changing-planet.

INDEX

Page numbers followed by (illus) indicate illustrations, and those followed by (fn) indicate footnotes on the same page.